AF618969

Synthesis Lectures on RF/Microwaves

This series publishes short books on theory, techniques and applications of guided wave and wireless technologies spanning the electromagnetic spectrum from RF/microwave through millimeter-waves and terahertz, including the aspects of materials, components, devices, circuits, modules, and systems which involve the generation, modulation, demodulation, control, transmission, sensing, and effects of electromagnetic signals.

Majid Pakdel

Understanding RF Systems

Components, Circuits, and Signal Processing

Majid Pakdel
MISCO (Mianeh Steel Complex)
Mianeh, Iran

ISSN 2150-0924 ISSN 2150-0932 (electronic)
Synthesis Lectures on RF/Microwaves
ISBN 978-3-032-19226-4 ISBN 978-3-032-19227-1 (eBook)
https://doi.org/10.1007/978-3-032-19227-1

This Springer imprint is published by the registered company Springer Nature Switzerland AG
The registered company address is: Gewerbestrasse 11, 6330 Cham, Switzerland

Preface

In the realm of modern communication, radio frequency (RF) technology plays an indispensable role, serving as the backbone for a myriad of applications ranging from traditional broadcasting to cutting-edge wireless communication systems. With the rapid evolution of technology and the increasing demand for efficient and reliable RF systems, understanding the fundamentals of RF components and circuits has never been more crucial. "Understanding RF Systems: Components, Circuits, and Signal Processing" aims to provide a comprehensive overview of the principles and practices that govern the design and operation of RF systems. This book is structured to cater to a diverse audience, including students, engineers, and enthusiasts eager to deepen their knowledge of RF technologies. It encompasses foundational concepts, detailed discussions of modulation techniques, and practical applications of RF components, ensuring that readers gain both theoretical insights and practical skills. Starting with an introduction to frequency and wavelength, the book progresses through essential topics such as amplitude modulation, mixer operations, and RF filter design. Each chapter is meticulously crafted to build upon the previous one, guiding readers through the intricacies of RF design methodologies. The inclusion of modern topics, such as software-defined radios (SDRs) and design using DeepSeek with Python, exemplifies the book's commitment to staying relevant in an ever-changing technological landscape. The chapters delve into critical components such as oscillators, amplifiers, antennas, and impedance matching, providing a thorough understanding of their roles in RF circuit design. Practical examples and illustrations are interspersed throughout the text, offering visual aids that enhance comprehension. Furthermore, the book includes a dedicated section on scattering parameters and the Smith chart, which are vital tools for RF engineers. Whether you are embarking on your journey into RF technology or looking to refine your expertise, this book serves as a valuable resource to guide you along the way. Your passion for RF technology is reflected in the pages that follow. Welcome to the exploration of RF components and circuits—may your journey be enlightening and inspiring.

Mianeh, Iran Majid Pakdel

Contents

Introduction

1

Contents

1.1 Frequency and Wavelength

The radio frequency (RF) signals travel as waves. The wavelength 'λ' is the distance traveled in one cycle as shown in Fig. 1.1.

The frequency 'f' of a signal is the number of cycles per second. The speed 'v' of a signal depends upon the medium and we have the following equation.

$$\lambda = \frac{v}{f} \tag{1.1}$$

In free space, the RF signal travels with the speed of light. If the RF signal travels under the water its speed will be decreased with respect to free space.

M. Pakdel, *Understanding RF Systems*, Synthesis Lectures on RF/Microwaves,
https://doi.org/10.1007/978-3-032-19227-1_1

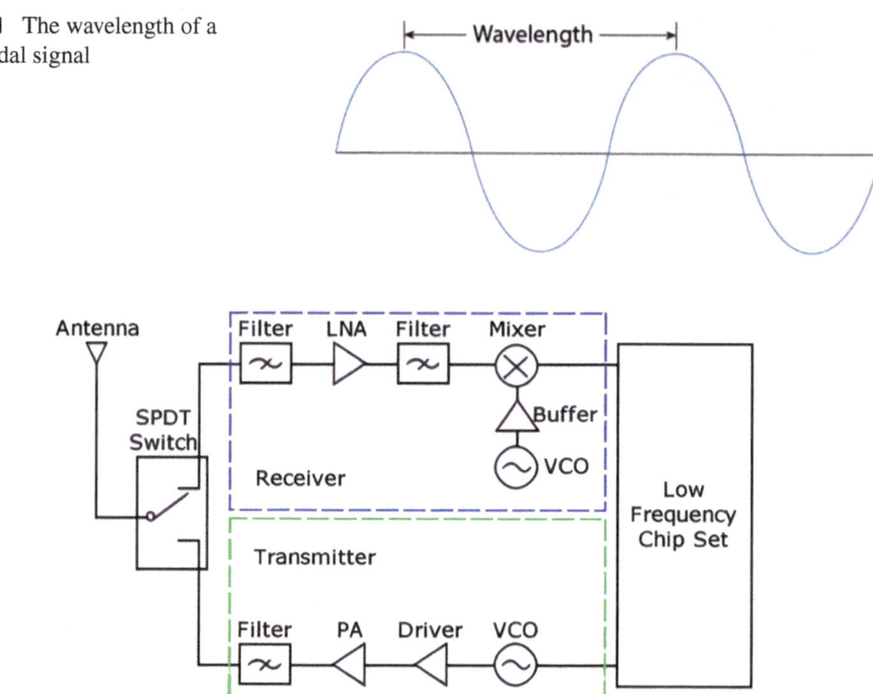

Fig. 1.1 The wavelength of a sinusoidal signal

Fig. 1.2 The two forms of RF signal

1.2 The RF Signal Frequency

The RF signals have traditionally defined frequencies from a few kilo hertz (kHz) to roughly one giga hertz (1 GHz). If one considers microwave frequencies as RF, this range extends to 300 GHz. The RF signal can be in one of two forms as below:

- As a current in conductor (transmitter and receiver sections in Fig. 1.2).
- As invisible electromagnetic signal on air (antenna section in Fig. 1.2).

1.3 Modulation/Demodulation for RF Communication

The most important procedures that the RF communication is possible are called modulation and demodulation. The RF transmitter performs the modulation while the RF receiver performs the demodulation as depicted in Fig. 1.3.

To understand clearly these concepts let us consider an example. Suppose a person is talking in the microphone in Fig. 1.3 and the microphone is converting the sound wave of that person to an electrical voice signal. The electrical signal has certain voltage which is

Fig. 1.3 The RF transmitter and receiver

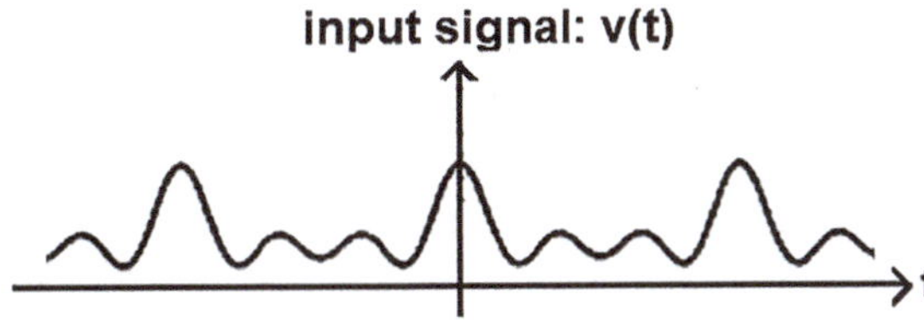

Fig. 1.4 The waveform of an electrical voice signal

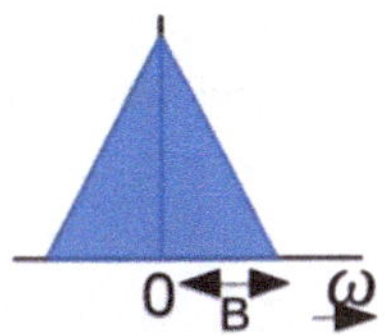

Fig. 1.5 The Fourier transform of input signal (baseband signal)

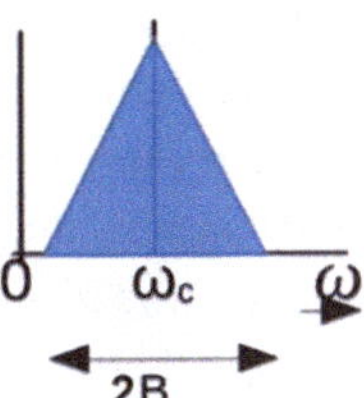

Fig. 1.6 The passband signal

a function of time. Since that is a function of time so we would call it as the time domain electrical signal as illustrated in Fig. 1.4.

If we want to know what frequencies there are in the electrical voice signal in Fig. 1.4, we can find them by taking the Fourier transform of that signal. So, we get the signal as a function of frequency and we can observe what frequencies there are in the signal as well as the magnitude of those frequencies as shown in Fig. 1.5.

So, the plot in Fig. 1.5 is a function of frequency and it will be centered in zero hertz (0 Hz). Therefore, it is called the baseband signal. The bandwidth (B) of that signal is from 0 Hz to the maximum frequency of the signal. The baseband signal is then given to the RF transmitter that modulates the signal and during the modulation procedure the baseband signal is moved to the passband signal. What is done by moving or shifting the signal to the passband is that now the signal is located at the carrier frequency ($\omega_c = 2\pi f_c$) as depicted in Fig. 1.6.

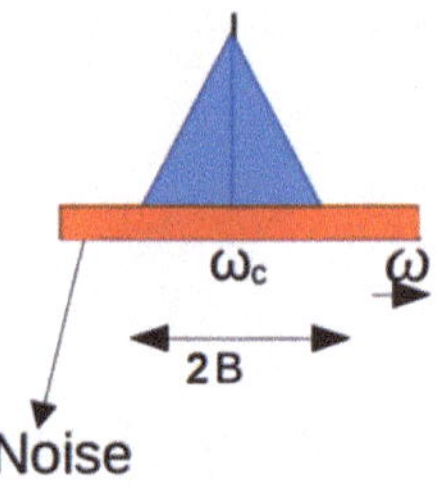

Fig. 1.7 The added wireless channel noise

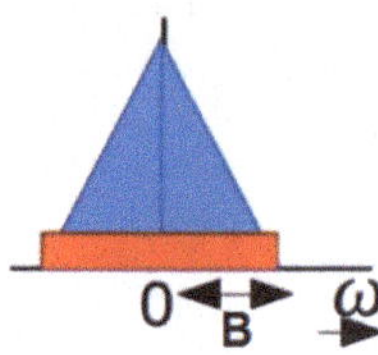

Fig. 1.8 Shifting the signal from passband to the baseband

Now, the signal is centered in the carrier frequency (ωc) and also the bandwidth of passband signal is twice of the baseband signal. Then, the baseband signal is transmitted through the antenna and it is converted to the electromagnetic wave and it would travel through the wireless channel. However, during this procedure the noise of channel would be added to the signal. So, the signal as illustrated in Fig. 1.7 is received by the receiver antenna.

Now, the RF receiver does the inverse procedure (demodulation), that means the receiver demodulates that signal in Fig. 1.7 and the signal would be moved from carrier frequency (ω_c) to the center frequency of 0 Hz. So, we again shift the signal from passband to the baseband as shown in Fig. 1.8.

Then, the signal in Fig. 1.8 is given to the speaker and the speaker can play this signal, however there would be a small noise in that signal. If the noise is small the sound quality in the speaker will be acceptable but if the noise is very large the sound quality will be not acceptable [1].

1.4 The Modulation

The modulation is the process by which some characteristics of a carrier wave is varied in accordance with the message signal. For modulation, according to message we can change the carrier's amplitude, frequency or phase. The carrier is a sinusoidal wave with the following equation.

$$c(A,\omega_c,\theta_0) = A\cos(\omega_c t + \theta_0) \tag{1.2}$$

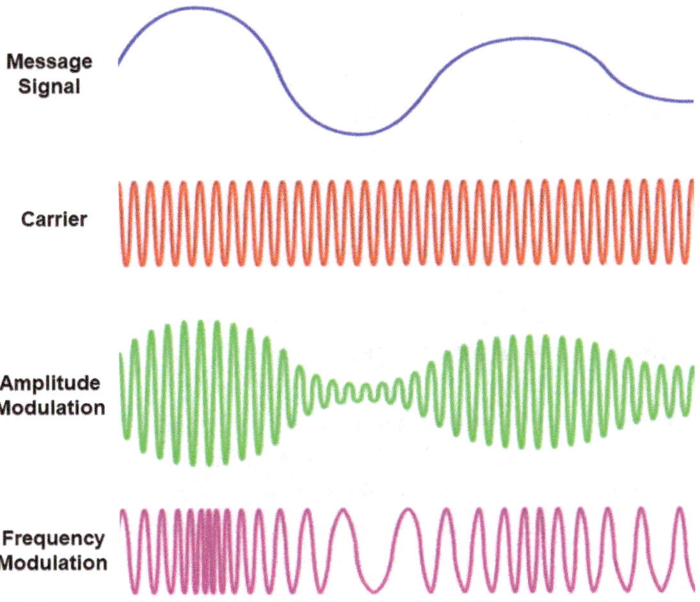

Fig. 1.9 The amplitude and frequency modulation

The carrier has a certain amplitude, a certain frequency, and a certain phase. During the modulation procedure, depending on the type of modulation that we are using, either we vary the amplitude of carrier in accordance to the message signal, or we vary the frequency of carrier in accordance to the message signal, or we vary the phase of carrier in corresponding to the message signal. For example, in the case of amplitude modulation, we vary the amplitude of carrier in corresponding to the message signal. So, when the amplitude of the message is increasing, we can view that the amplitude of the carrier is also increasing and when the message amplitude is maximum, the carrier amplitude is also maximum and vice versa. In the case of frequency modulation, we vary the frequency corresponding to the amplitude of the message signal. So, when the amplitude of the message is increasing, we can see that the frequency of the carrier is also increasing and when the message amplitude is maximum, the carrier frequency is also maximum and vice versa as depicted in Fig. 1.9.

The phase modulation is not shown in Fig. 1.9. In the case of phase modulation, we vary the phase of the carrier in accordance of the message signal [1].

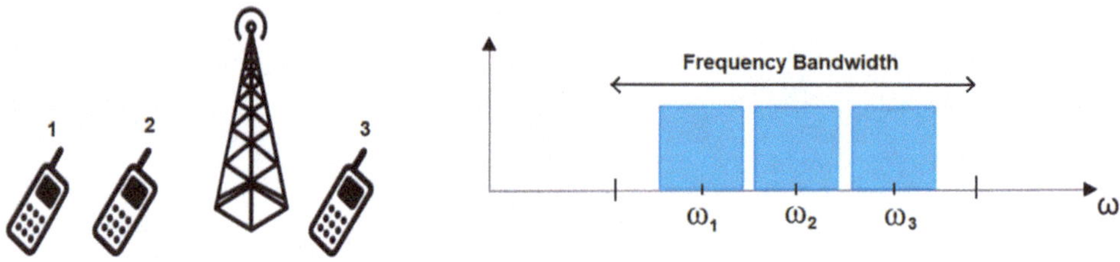

Fig. 1.10 The frequency division multiplexing

1.5 The Modulation Advantages

The modulation causes a shift in the center frequency of a signal and it is moved from zero frequency to the carrier frequency. The first advantage of modulation is that the size of antenna for transmitting and receiving of RF signals is inversely proportional to the frequency of the signal [2]. The second advantage of modulation is that it allows multiple signals with the same frequency bandwidth to be transmitted simultaneously (frequency division multiplexing). For example, the Fig. 1.10 illustrated the frequency bandwidth which has been assigned via the mobile operator, and there is a tower for that mobile operator and three mobiles want to communicate with the mobile network using the mobile operator tower. If they use the same frequency, then there will interfere with one another. So, in order to avoid that the mobile station 1 would send or would modulate its signal at the carrier frequency of ω1, and will send it to the tower and the mobile station 2 will use the carrier frequency of ω2, and will send it to the tower, and the mobile station 3 will use the carrier frequency of ω3 in order to send its signal to the tower. Now the carrier frequencies have been located in different frequencies and as a result they will not interfere with one another. So, in this way multiple signals can be transmitted within the given bandwidth and this procedure is called the frequency division multiplexing.

1.6 The Analog Modulation Types

The modulation has two main types which are analog modulation and digital modulation and first we want to focus on analog modulation. The analog modulation is further subdivided into three major types including amplitude modulation, phase modulation and frequency modulation. We have also quadrature amplitude modulation which is the combination of amplitude modulation and phase modulation. The block diagram of modulation types is shown in Fig. 1.11.

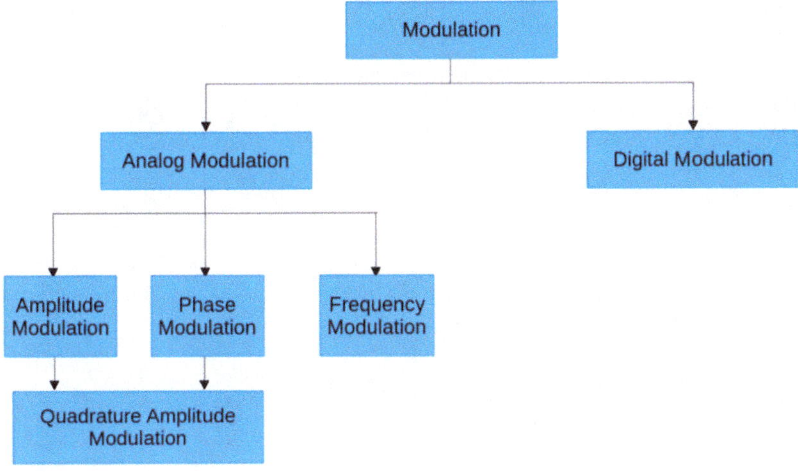

Fig. 1.11 The block diagram of analog modulation types

1.7 Conclusion

This introductory chapter has established the fundamental pillars upon which the entire edifice of RF system design is built. We began by defining the core physical concepts of frequency and wavelength, along with their governing relationship, laying the groundwork for understanding how RF signals propagate through different media.

The chapter then framed the central challenge of RF communication: bridging the gap between low-frequency information signals and efficient wireless transmission. We explored the indispensable roles of modulation at the transmitter and demodulation at the receiver, illustrating how these processes transform a baseband signal into a transmittable passband signal and recover it at the destination. This was complemented by a practical examination of the significant advantages modulation provides, most notably enabling practical antenna design and permitting multiple, simultaneous communications through frequency-division multiplexing.

Finally, we introduced the primary taxonomy of analog modulation—Amplitude Modulation (AM), Frequency Modulation (FM), and Phase Modulation (PM)—setting the stage for their detailed mathematical and circuit-level analysis in subsequent chapters. By concluding with a conceptual block diagram of analog modulation types, this chapter provides a clear map of the territory ahead.

In essence, Chap. 1 has equipped the reader with the essential language and high-level principles of RF systems. It has shown that at its heart, RF engineering is the art and science of intelligently manipulating a sinusoidal carrier wave to convey information reliably across space. With this conceptual foundation in place, we now proceed to delve into the specific components, circuits, and sophisticated signal processing techniques that bring these principles to life.

References

1. Tiwana M (2021) RF concepts, components and circuits for beginners. Udemy Inc., San Francisco, CA
2. Pozar DM (2011) Microwave engineering, 4th edn. Wiley, Hoboken, NJ

Amplitude Modulation 2

Contents

2.1 The Amplitude Modulation Scheme

In the amplitude modulation (AM) which is an analogue modulation scheme we can easily generate the amplitude modulation signal by multiplying the message signal to the carrier. So, if we have the message signal (m(t)), and the carrier (cos(ωct)) we can obtain the modulated signal by the following equation.

$$Modulated\ Signal = m(t)\cos(\omega_c t) \tag{2.1}$$

If we take the Fourier transform of original message signal, we can see that it is centered at zero frequency and it has the bandwidth of B. Also, if we take the Fourier transform of the modulated signal, we can see that the original message signal is now shifted to the carrier frequency and its bandwidth is twice (2B) of the original baseband message signal bandwidth as shown in Fig. 2.1 [1].

M. Pakdel, *Understanding RF Systems*, Synthesis Lectures on RF/Microwaves,
https://doi.org/10.1007/978-3-032-19227-1_2

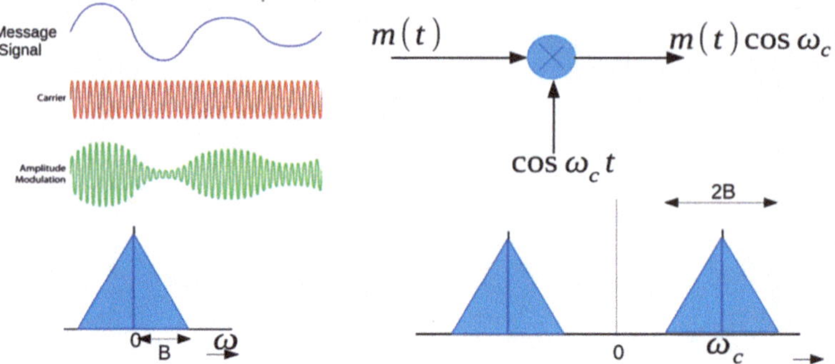

Fig. 2.1 The amplitude modulation scheme

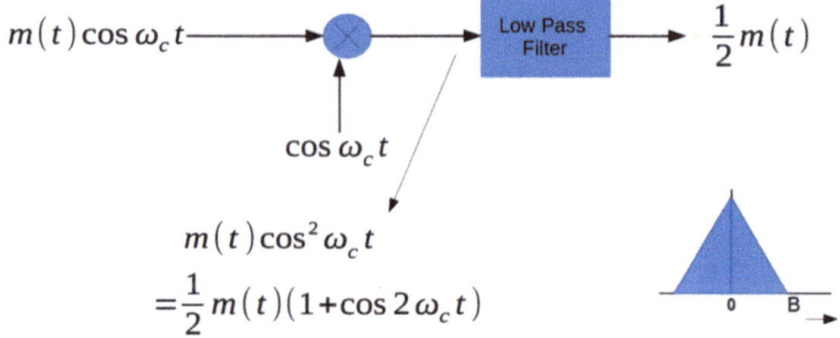

Fig. 2.2 Demodulation of AM signal

2.2 Demodulation of AM Signal

Now we can easily demodulate that amplitude modulated (AM) signal by multiplying it to the same carrier that we used at the transmitter. So, we will have the following equation.

$$\textit{Demodulated Signal} = m(t)\cos^2(\omega_c t) = \frac{1}{2}m(t)\big(1+\cos(2\omega_c t)\big) \tag{2.2}$$

There are two main components in the Eq. (2.2), one is the low frequency component which is centered at zero frequency and the other is the high frequency component that is located at $2\omega_c$. However, we do not want that high frequency component, so we will pass the signal through the lowpass filter and the lowpass filter will allow the low frequency component to pass but it will block the higher frequency component. That means the component $\cos(2\omega_c t)$ will be filter out and at the output we would get the low frequency component (1/2 m(t)) and the m(t) is the same message that we transmitted using the modulator and now we have recovered it and also it is located at center frequency of zero hertz (0 Hz) as depicted in Fig. 2.2.

For demodulation procedure, we apply the same carrier ($\cos(\omega_c t)$) that we used at the transmitter and we need exact carrier frequency and phase at the receiver for correct demodulation. So, for demodulation we need to recover the carrier and for recovery of the carrier, we use a device which is called as the phase lock loop (PLL). Also, one of the most popular PLLs that is used for carrier recovery is called as the Costas loop [1].

2.3 RF Transmitter Block Diagram

The major components of RF transmitter are illustrated in Fig. 2.3. The message signal that we want to transmit is first given to the amplifier and it increases the amplitude of the message signal. The oscillator is a device that generates the carrier. So, using the mixer we do multiply the message signal with the carrier and at the output of mixer we get the AM signal and then that signal is passed through a filter and that filter is a band-pass filter and that means the filter has been tuned to the carrier frequency of ω_c. The frequencies around the carrier frequency are allowed to pass while the other frequencies including noise frequencies are blocked by the filter. So, the noise is blocked and only the signal is allowed to pass. After that the signal is amplified and then it is transmitted using the antenna [2].

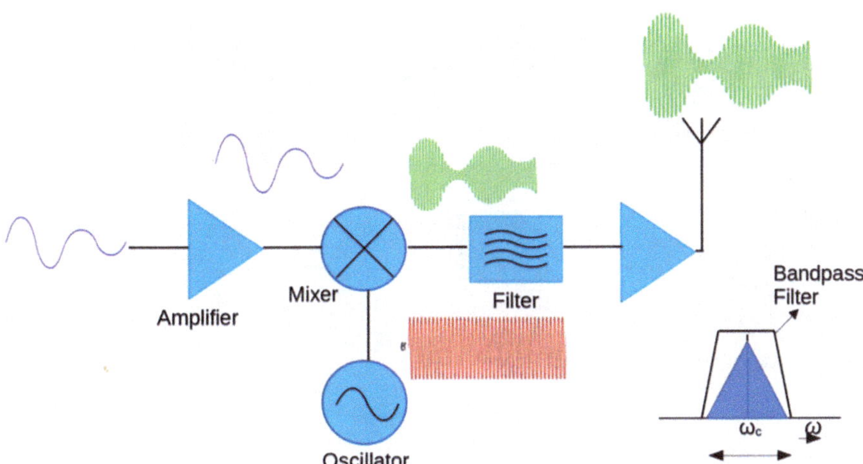

Fig. 2.3 The major components of RF transmitter

2.4 RF Receiver Block Diagram

The modulated signal that is received in the antenna is weak and it has noise in it. So, first of all we pass that signal through a bandpass filter which is tuned at the frequency of ω_c and it is the carrier frequency. So, that filter will allow the frequencies around the carrier frequency to pass through it while it would block the noises that are at the other frequencies. Then, that signal goes ahead to the low noise amplifier (LNA), you should note that the signal enters to the LNA is a weak signal, so it should amplify that weak signal and the noise that introduced by LNA in to the signal should be as low as possible. Then we have the PLL that works in conjunction with the oscillator in order to generate the carrier frequency and that carrier frequency has the same phase and frequency as the carrier frequency that is used at the transmitter. After that using the mixer, we would multiply the carrier with the signal and then we would pass it through the lowpass filter. So, we can demodulate the AM signal and that lowpass filter basically allows the signals that centered at zero frequency to pass through it and that is our desired signal. Also, we may need to amply it further, so we may use an amplifier to amplify it further as shown in Fig. 2.4 [2].

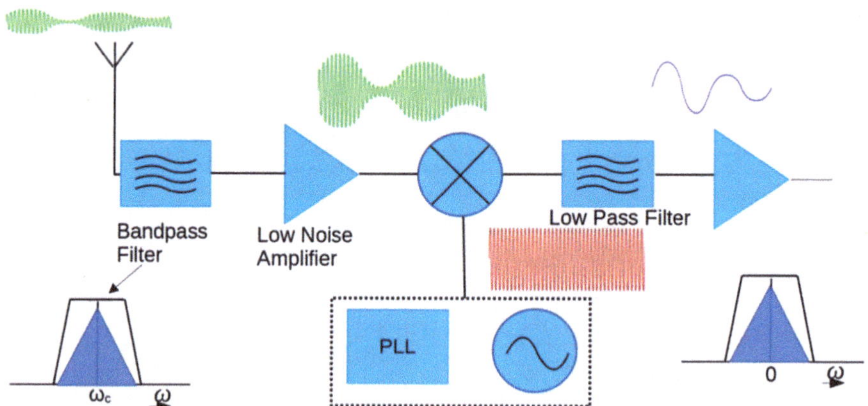

Fig. 2.4 The major components of RF receiver

2.5 Conclusion

This chapter has provided a comprehensive examination of Amplitude Modulation (AM), establishing it as a fundamental building block in the architecture of RF communication systems. Beginning with the core mathematical principle of multiplying a message signal by a carrier, we demonstrated how AM achieves the essential task of shifting a baseband signal to a higher frequency, thereby enabling its efficient radiation via an antenna. The spectral transformation from a baseband bandwidth B to a passband bandwidth 2B was clearly illustrated, reinforcing the concept of spectrum utilization introduced in Chap. 1.

A key focus of the chapter was the complete signal processing chain of AM. We detailed the coherent demodulation process, showing that recovery of the original message is contingent upon multiplying the received signal with an exact replica of the transmitter's carrier. This underscores a critical practical challenge in receiver design: the need for precise carrier frequency and phase synchronization, typically addressed by circuits like the Phase-Locked Loop (PLL).

Finally, the theoretical concepts were grounded in practical hardware implementation. Through detailed block diagrams, the functions of each major component in an AM transmitter and receiver—from oscillators and mixers to amplifiers and filters—were explained. Special attention was given to critical components like the Low-Noise Amplifier (LNA) and the PLL, which are vital for managing signal integrity and ensuring robust demodulation in the presence of noise.

In summary, Chap. 2 has successfully bridged the abstract mathematical formulation of AM with its tangible circuit-level realization. By understanding AM's operation, strengths, and inherent requirements for coherent reception, a solid foundation is now laid for exploring more spectrally efficient and noise-resistant modulation schemes, such as Frequency Modulation (FM) and Phase Modulation (PM), in subsequent chapters.

References

1. Tiwana M (2021) RF concepts, components and circuits for beginners. Udemy Inc., San Francisco, CA
2. Pozar DM (2011) Microwave engineering, 4th edn. John Wiley & Sons, Hoboken, NJ

Analog and Digital Modulation Schemes 3

Contents

3.1 Analog Quadrature Amplitude Modulation

Analog QAMs are used to carry multiple signals on a single carrier. It is similar to AM but with two carrier signals at the same frequency and 90 degrees out of phase. The AM signal occupies twice the bandwidth required for the baseband signal as demonstrated in Fig. 2.1. This disadvantage can be overcome in the QAM by transmitting two signals within same bandwidth (BW) using the same carrier by the following equation.

M. Pakdel, *Understanding RF Systems*, Synthesis Lectures on RF/Microwaves,
https://doi.org/10.1007/978-3-032-19227-1_3

$$\varphi_{QAM} = m_1(t)\cos(\omega_c t) + m_2(t)\sin(\omega_c t) \tag{3.1}$$

The $m_1(t)$ is called the in-phase (I) component and $m_2(t)$ is called the quadrature (Q) component, also the φ_{QAM} is called the quadrature amplitude modulated signal. The baseband signals $m_1(t)$ and $m_2(t)$ will not interfere with each other because the carrier signals $\cos(\omega_c t)$ and $\sin(\omega_c t)$ are orthogonal with each other which is expressed by the equation below.

$$\int_0^T \sin(\omega_c t)\cos(\omega_c t)\,dt = 0 \tag{3.2}$$

The modulation and demodulation diagram of analog QAM is shown in Fig. 3.1 [2].

To recover $m_1(t)$, from Fig. 3.1, we can write the following equation.

$$x_1(t) = 2\big(m_1(t)\cos(\omega_c t) + m_2(t)\sin(\omega_c t)\big)\cos(\omega_c t) \tag{3.3}$$

We use the equations below to simplify further the Eq. (3.3).

$$\cos^2(\omega_c t) = \frac{1+\cos(2\omega_c t)}{2},\ \sin(\omega_c t)\cos(\omega_c t) = \frac{\sin(2\omega_c t)}{2} \tag{3.4}$$

So, using the Eqs. (3.3) and (3.4), we will have the equation below.

$$x_1(t) = m_1(t) + m_1(t)\cos(2\omega_c t) + m_2(t)\sin(2\omega_c t) \tag{3.5}$$

Then, we recover the baseband signal $m_1(t)$ using a lowpass filter (LPF). To recover $m_2(t)$, from Fig. 3.1, we can write the following equation.

$$x_2(t) = 2\big(m_1(t)\cos(\omega_c t) + m_2(t)\sin(\omega_c t)\big)\sin(\omega_c t) \tag{3.6}$$

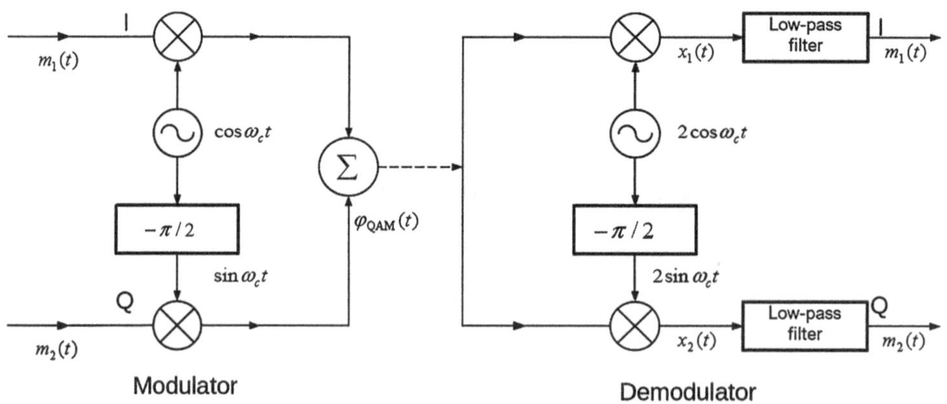

Fig. 3.1 The modulation and demodulation diagram of analog QAM

We use the equations below to simplify further the Eq. (3.6).

$$\sin^2(\omega_c t) = \frac{1-\cos(2\omega_c t)}{2}, \ \sin(\omega_c t)\cos(\omega_c t) = \frac{\sin(2\omega_c t)}{2} \tag{3.7}$$

So, using the Eqs. (3.6) and (3.7), we will have the equation below.

$$x_2(t) = m_2(t) + m_1(t)\sin(2\omega_c t) - m_2(t)\cos(2\omega_c t) \tag{3.8}$$

Then, we recover the baseband signal $m_2(t)$ using a lowpass filter (LPF).

3.2 The Digital Modulation Types

We demonstrated the types of analog modulation schemes in Fig 1.11. In this section, we are going to explain the types of digital modulation schemes as depicted in Fig. 3.2 [1].

In the digital modulation schemes, you have the amplitude shift keying (ASK) which is the digital version of amplitude modulation (AM), you have phase shift keying (PSK) which is the digital version of phase modulation (PM), you have frequency shift keying (FSK) which is the digital version of frequency modulation (FM), and the quadrature amplitude modulation (QAM) which is the combination of ASK with PSK, however that QAM is in digital domain.

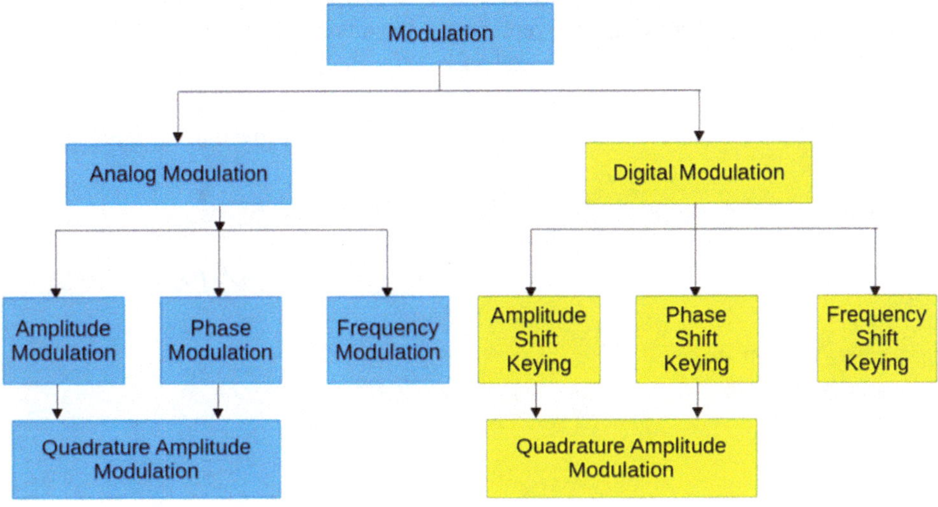

Fig. 3.2 The block diagram of modulation types

3.3 The Binary Amplitude Shift Keying (ASK)

In this section we are going to explain the digital modulation scheme of binary ASK. In binary ASK, the one and zero levels are represented by two carrier amplitude levels and it is also called the on-off keying (OOK). Suppose we want to send the bits 011010 using the binary ASK, first of all we need to convert those bits to the symbols, so we assign a zero-voltage level for bit 0 and some voltage level for bit 1 and we will have a waveform of symbols and then we will multiply that waveform by $\cos(\omega ct)$ and we will get the binary ASK modulated signal at the output as illustrated in Fig. 3.3. As you can see in Fig. 3.3, for bit 0 we do not send anything however for bit 1 we send the carrier.

There is another important concept which is called the constellation diagram which is used for different modulation schemes. The binary ASK modulation scheme can be represented by the constellation diagram as shown in Fig. 3.4.

Since the symbols in Fig. 3.3 have been multiplied by $\cos(\omega_c t)$, those symbols lie on the axis of $\cos(\omega_c t)$, or in the other words those symbols lie on the I (in-phase) axis. For the symbol 0 we are not sending anything for the 0 and it has zero energy. So, we can represent the symbol 0 by a constellation point at the origin. For the symbol 1, we have assigned some voltage for it and we are sending carrier for symbol 1 and it has some energy. So, the symbol 1 is represented by a constellation point there and the distance from that constellation point to the origin will be the root of the energy of that symbol in its modulated form. Since, we are not using the $\sin(\omega_c t)$ for this binary ASK modulation, we will not have any symbol along with the axis of $\sin(\omega_c t)$, or in other words, we can say that there is no quadrature component (Q) in the binary ASK. Also, since we have two symbols in the constellation diagram that means we send 1 bit/symbol i.e., one symbol is used to represent 0 and the other symbol is used to represent 1 as you can view in Fig. 3.4.

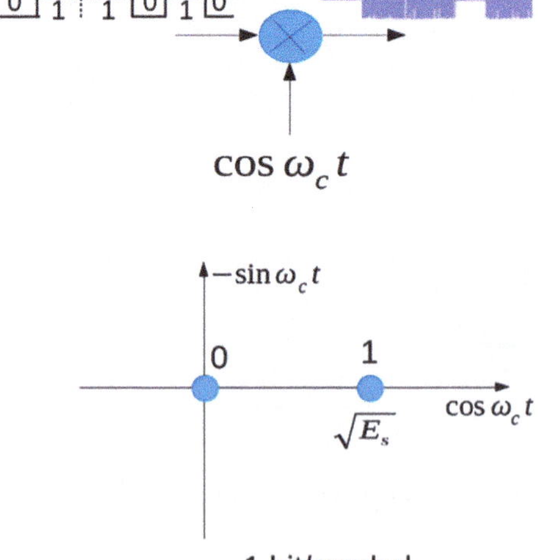

Fig. 3.3 The binary ASK modulation

Fig. 3.4 The constellation diagram for binary ASK modulation scheme

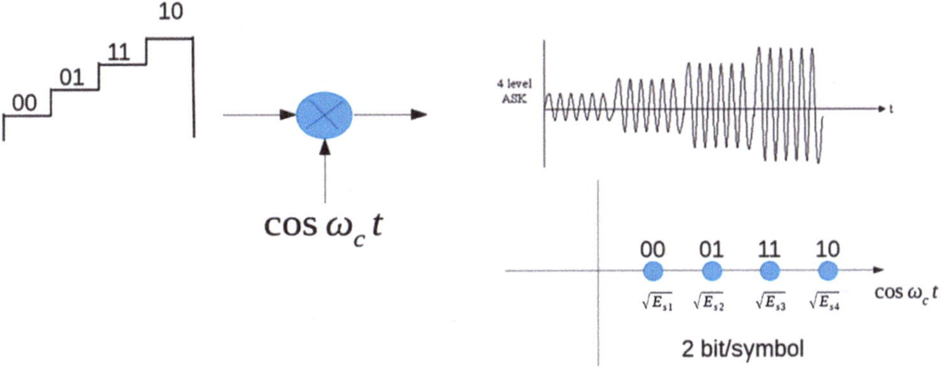

Fig. 3.5 The 4-ASK digital modulation scheme

3.4 The 4-ASK Digital Modulation

In this section we are going to explain the 4-ASK digital modulation scheme. Here, 4 tells us there are 4 symbols in this modulation scheme, in other words there are 4 types of symbols in this modulation scheme and corresponding to these 4 symbol types, there are 4 levels of carrier between which the carrier is switched. For example, the 00 is being represented by a symbol which has a certain voltage level, the 01 Is being represented by a symbol that has another voltage level, the 11 is being represented by a symbol that has another different voltage level, and finally the 10 is being represented by a symbol that has also another different voltage level and then we will multiply those symbols by the carrier and as a result we will get a signal where the carrier is switched between 4 levels according to the corresponding symbol. Also, those four symbols lie along the axis of $\cos(\omega_c t)$ or the I axis in the constellation diagram and their locations will be determined by the energy of those symbols. Since, we have 4 symbols, we can send 2 bits by each symbol as depicted in Fig. 3.5.

3.5 The 4-ASK Demodulation

In this section we will explain the demodulation of 4-ASK signal. So, we are going to use the same procedure as we used for the demodulation of the AM signal as illustrated in Fig. 3.6.

As you can see in Fig. 3.6, we multiply the received 4-ASK signal by the carrier signal and then pass it through the lowpass filter and after that we recover the baseband symbols. Then we sample those baseband symbols and we apply the thresholding function on those sampled symbols. In the 4-ASK constellation diagram, there are 4 constellation points and also there are 3 thresholds within these constellation points as shown in Fig. 3.7.

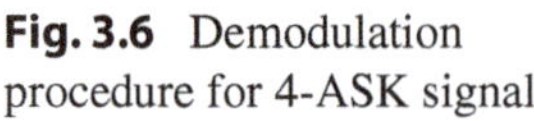

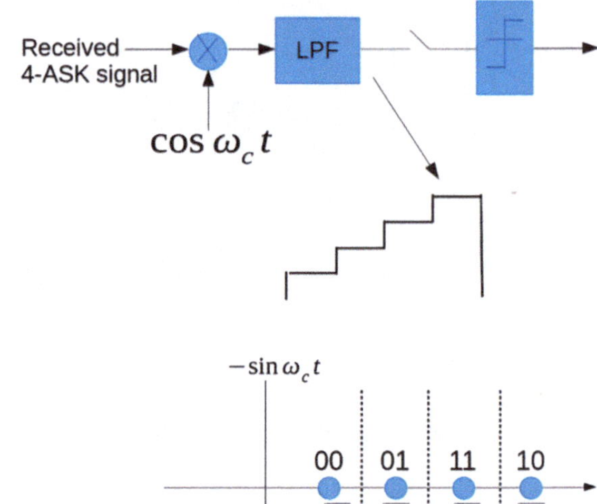

Fig. 3.6 Demodulation procedure for 4-ASK signal

Fig. 3.7 The 4-ASK constellation diagram with 3 thresholds

So, when we sample the baseband symbols and that symbol lies at certain point in Fig. 3.7, which is between 2 thresholds, then we can obtain the corresponding 2 bits were transmitted. So, in this way we demodulate the 4-ASK signal and we recover the bits that were transmitted by the transmitter.

3.6 The Binary Frequency Shift Keying (BFSK)

In this section we are going to discuss the binary frequency shift keying (BFSK). In BFSK, the instantaneous carrier frequency is switched between 2 values. The data bits select a carrier at one of two frequencies for example 0 is being represented by the carrier frequency of ω_1 and 1 is being represented by the carrier frequency of ω_2. So, when we are sending 0, we will send the carrier frequency of ω_1 and when we are sending 1, we will send the carrier frequency of ω_2. So, depending upon whether we are sending 0 or 1, we will switch the carrier frequencies between the ω_1 and ω_2 and this type of BFSK modulation can be very easily generated using the configuration as depicted in Fig. 3.8.

In Fig. 3.8, the oscillator 1 is generating the carrier frequency of ω_1, and oscillator 2 is generating the carrier frequency of ω_2. Also, we have a switch and using that switch we select whether we are sending the ω_1 or whether we are sending the ω_2 and that switch is operated using the bits that we are sending. So, if we are sending 0, the switch will be in position of ω_1, and it would send the ω_1 carrier frequency, also, if we are sending 1, the switch will be in position of ω_2, and it would send the ω_2 carrier frequency. So, the data is encoded in the frequency and until recently, the BFSK has been the most widely used form

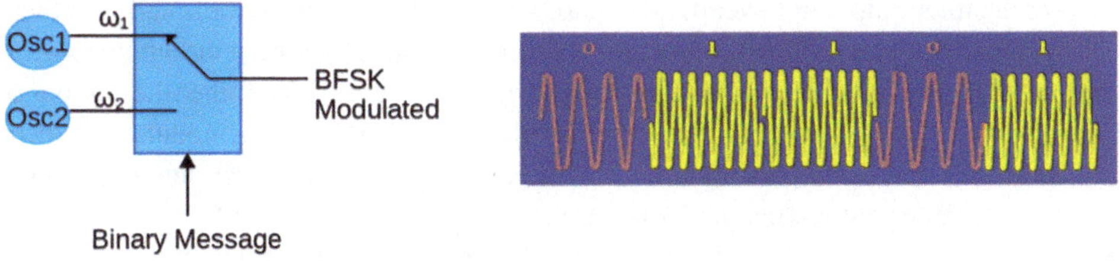

Fig. 3.8 The BFSK modulation scheme

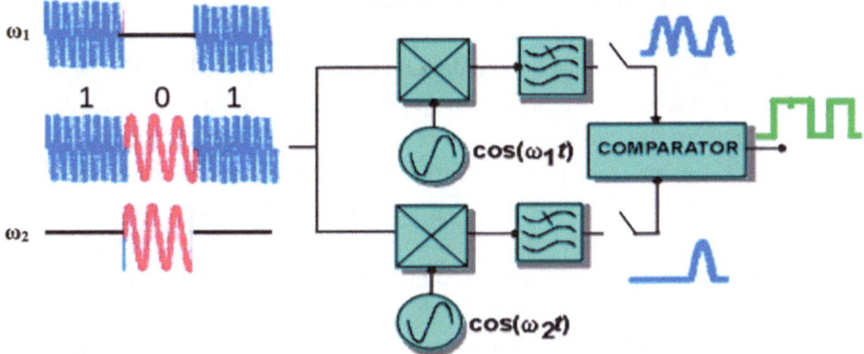

Fig. 3.9 The BFSK demodulation scheme

of digital modulation because it is very easy to generate (modulate) and also it is very easy to detect (demodulate). Also, the BFSK is insensitive to amplitude fluctuations in the channel. Since, most of the time the noise changes the amplitude of the transmitted signal. However, in this case the information is being carried as frequency and there is a very little effect on the frequency, so the FSK is more robust to the noise.

3.7 The BFSK Demodulation

We can easily demodulate the BFSK modulated signal because the BFSK signal is a combination of two binary ASK (BASK) signals. For example, in Fig. 3.9, we have a BFSK signal where 1 is being represented by the carrier frequency of ω_1 and 0 is being represented by the carrier frequency of ω_2. So, that BFSK signal is the combination of two BASK signals. In order to receive these two BASK's signals, we would need two BASK demodulators that would be tuned to two different carrier frequencies and these two BASK demodulators would be connected in parallel. So, one of demodulators has the carrier frequency of ω_1 while the other has the carrier frequency of ω_2. After multiplying by the carrier frequencies, we pass the resulted signals through the lowpass filters. When we sending 1 with carrier frequency of ω_1, the filter output of upper demodulator will be high, however the filter output of lower demodulator will be zero, so if we sample and pass them to the comparator, since upper filter output is greater that the lower filter output that means 1 was sent. However, when we sending 0 with carrier frequency of ω_2, the filter output of upper demodulator will be zero, however the filter output of lower demodulator will be high, so if we sample and pass them to the comparator, since lower filter output is greater that the upper filter output that means 0 was sent as illustrated in Fig. 3.9.

So, in this way we can demodulate the BFSK signal as two BASK signals.

3.8 The Binary Phase Shift Keying (BPSK)

In this section we are going to discuss the digital modulation scheme of binary phase shift keying (BPSK). In the case of phase shift keying, the information is contained in the instantaneous phase of the carrier. For example, in the BPSK we are sending one bit per symbol and 1 is being represented by the carrier that has the 0-degree phase and 0 is being represented by the carrier which has 180-degree phase. So, whenever we have the transition from 1 to 0 or 0 to 1, we have the phase shift of 180 degrees. The BPSK signal is generated as follows, first of all we convert bits to the symbol and we represent 1 with the symbol that has positive voltage level and 0 with the symbol that has negative voltage level to get the waveform of symbols and then we multiply these symbols with the carrier which generates the modulated symbols. We can see that the symbols for 1 and 0 have the same amplitude but they have 180-degree phase shift in the phase or in other words their polarity is opposite to one another, so the constellation diagram the energy of value of the constellation points would be the same since the carrier amplitude is the same as shown in Fig. 3.10.

Since we are only using the carrier of $\cos(\omega_c t)$, the constellation points lie only on the axis of $\cos(\omega_c t)$ and we are not using the quadrature component ($\sin(\omega_c t)$) in the case of BPSK.

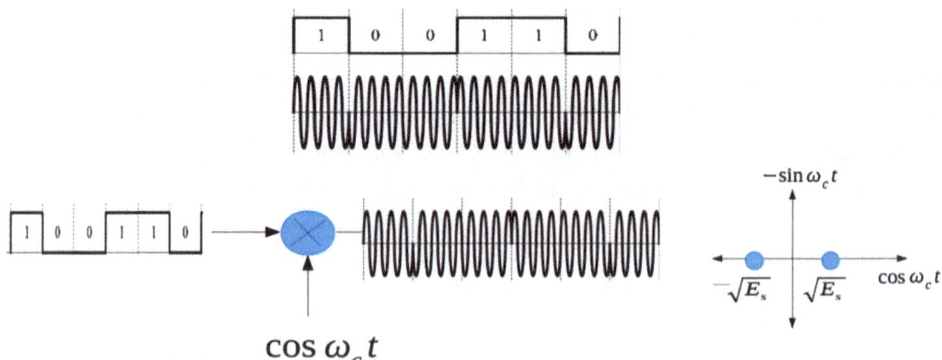

Fig. 3.10 The BPSK modulation scheme

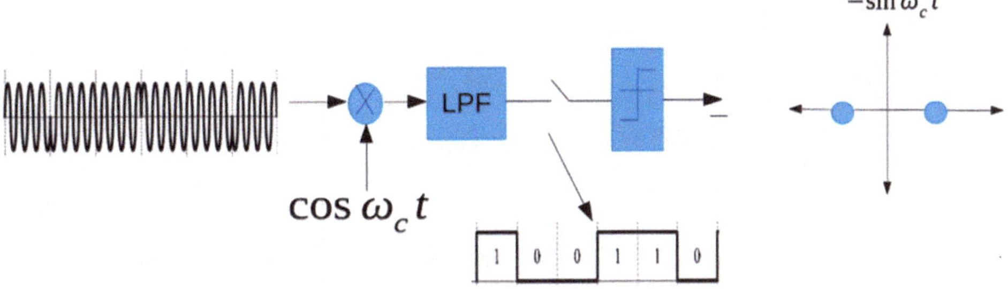

Fig. 3.11 The BFSK demodulation scheme

3.9 The BPSK Demodulation

The demodulation of BPSK can be done in the same way as we demodulated the AM signal. The received BPSK signal is multiplied by the carrier then it is passed through the LPF and as a result we get the baseband symbols. After that we sample those symbols and we do sampling in the middle of symbols and then we apply the thresholding function on those sampled symbols and on the constellation diagram since we have two constellation points the threshold is in between of them and that is how we recover the original transmitted bits as depicted in Fig. 3.11.

3.10 The M-Phase Shift Keying (MPSK)

In this section we will discuss the concept of MPSK, where M represents the number of symbols. When M = 2, we have only 2 symbols in the constellation diagram and one of the symbols has the phase of 0 degree and the second symbol has the phase of 180 degrees and also since there are only 2 symbols in the BPSK signal so we are sending 1 bit per symbol. For M = 4, we have 4 symbols in the constellation diagram, and the first symbol has the phase of 0 degree, the second symbol has the phase of 90 degrees, the third symbol has the phase of 180 degrees and the fourth symbol has the phase of 270 degrees. Since there are 4 symbols so each symbol represents 2 bits. The first symbol represents the "00" bits, the second symbol represents the "01" bits, the third symbol represents the "11" bits, and the fourth symbol represents the "10" bits. We use grey coding between adjacent symbols to reduce the bit error, which there is one bit change between 2 symbols so in the error condition in adjacent symbols, we will have only one bit error instead of two-bit error. So, the advantage of using the grey coding is that it reduces the bit error between the adjacent symbols and this error occurs due to the noise. In the case of M = 8, we have 8 symbols that means we have 8 phases corresponding to those symbols on the constellation diagram and for those 8 symbols we are sending 3 bits per symbol as illustrated in Fig. 3.12.

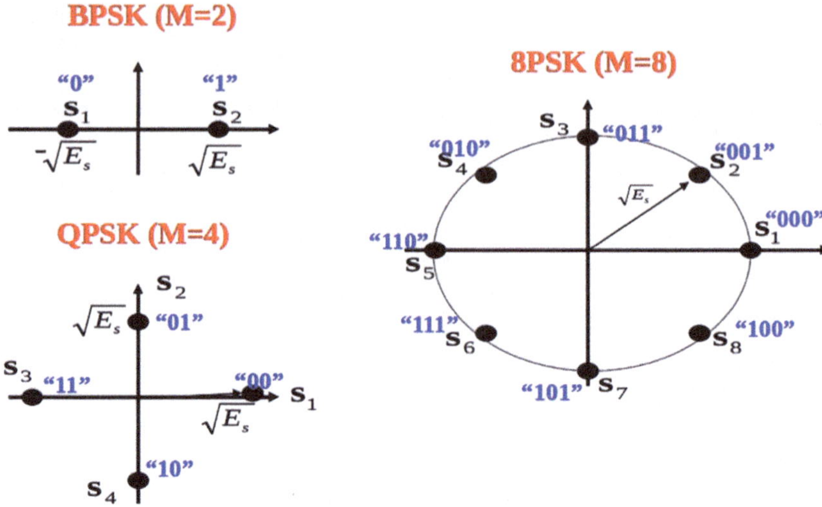

Fig. 3.12 The constellation diagrams for MPSK

Now, the important point about the MPSK is that the information is carried in the phase of the symbols but each of the symbols has the same amplitude that means each of the symbols has the same energy and also that means each of those constellation points lie on a circle, because all of them have the same energy.

3.11 The Quadrature Phase Shift Keying (QPSK) Modulation

In this section we will discuss about how the QPSK modulation is done. In QPSK, M = 4, and that means there are 4 symbols and in each of 4 symbols there are an I component and a Q component. For example, for the symbol denoted by "00", the I and Q components are both equal to 1, also for the symbol denoted by "01", the I component is equal to −1 and the Q components is equal to 1 and so on. The QPSK signal is composed of two BPSK signals. The first BPSK signal is along the I axis and the second BPSK signal is along the Q axis. We suppose that we have the "00" bits in the input of the modulator, and also, we suppose that the first bit belongs to the I-BPSK signal and the second bit belongs to the Q-BPSK signal, so the first bit would be sent to the I-channel and it would be mapped to the I-level and its I-level is 1 and it would be modulated using the $\cos(\omega_c t)$ carrier. The second bit belongs to the Q-BPSK signal and it would be sent to the Q-channel and it would be mapped to the Q-level and its Q-level is 1 and it would be modulated using the -$\sin(\omega_c t)$ carrier, and then we would add two signals and pass it through the bandpass filter so we will get the QPSK modulated signal against to the "00" bits. Similarly, if the next symbol is "10", the first bit (0) is sent to the I-channel and we know that the 0 is mapped to the 1 and then it would be modulated using the $\cos(\omega_c t)$ carrier, and the second bit (1) is

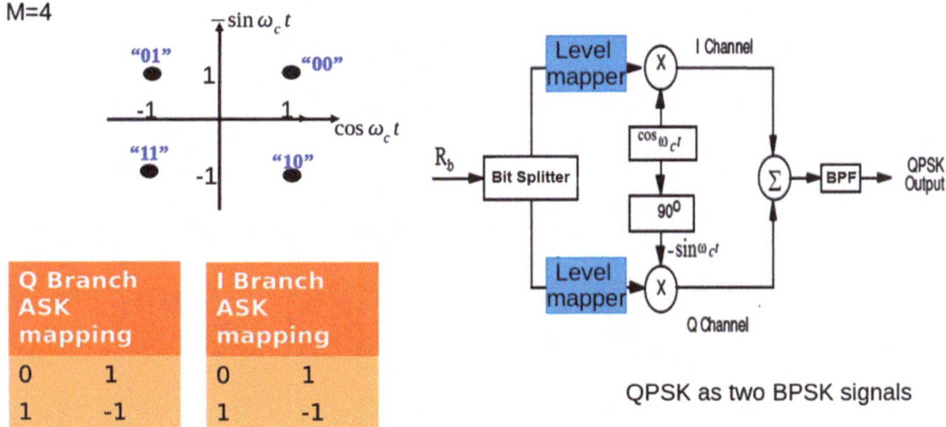

Fig. 3.13 The QPSK modulation scheme

sent to the Q-channel and we know that the 1 is mapped to the −1 and then it would be modulated using the -sin(ω_ct) carrier and then we would add those two signals and pass it through BPF to generate the QPSK signal against the "10" bits. So, we can see that we are generating the QPSK signal as a combination of two BPSK signals as shown in Fig. 3.13.

3.12 The QPSK Demodulation

We know that the QPSK signal can be generated as the sum of two BPSK signals, in a similar manner, the QPSK signal can be demodulated using two BPSK demodulators. One of the BPSK demodulators can be used to demodulate the I component of the received QPSK signal and the other BPSK demodulator is used to receive the Q component of the received QPSK signal. We suppose a symbol has been received by the receiver, now in the I branch, this symbol will be multiplied by the cos(ω_ct) carrier and then it would be passed to the lowpass filter (LPF) and as a result we will get the corresponding I component in the I axis at the output of the LPF. Then the I component passes through the binary threshold and that threshold would be on the I-BPSK signal (Q axis) and so the thresholder decides what I-level was transmitted. After that the output of binary threshold will be the I-level and it would be mapped to the corresponding bit by the level to bit mapper. Similarly, when the received symbol is multiplied by the -sin(ω_ct) carrier and it is passed to the LPF, we would get the Q component of that symbol and we will get the corresponding Q-level and that would be the input to the binary thresholder (I axis) and so the thresholder decides what Q-level was transmitted. After that the output of binary threshold will be the Q-level and it would be mapped to the corresponding bit by the level to bit mapper. Finally, we would convert those bits into the serial bits by the parallel to serial block and we can get the actual bits that was transmitted as depicted in Fig. 3.14.

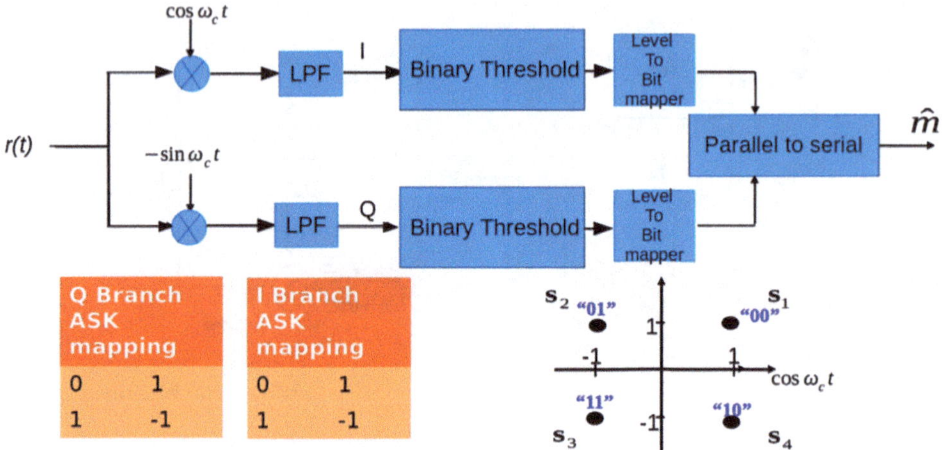

Fig. 3.14 The QPSK demodulation scheme

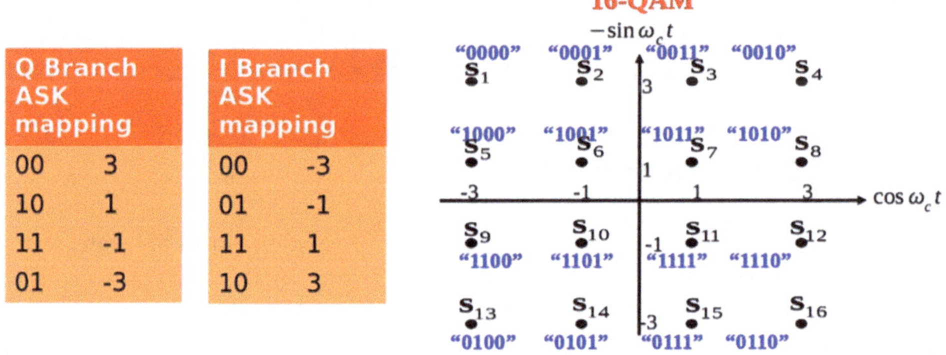

Fig. 3.15 The 16QAM constellation diagram

3.13 The M-Quadrature Amplitude Modulation

In the MQAM, we not only vary the phase but also, we vary the amplitude. So, we will have different amplitude and phase for symbols. In order to generate the MQAM modulation, we can generate it using two $\sqrt{M}$ -ASK signals. For example, in order to generate the 16QAM signal, we would need two 4ASK signals, The first 4ASK signal is 4 I-level symbols along with the I axis and the second 4ASK signal is 4 Q-level symbols along with the Q axis and using those two 4ASK signals, we can generate any symbol of the 16QAM modulation and since there are 16 symbols so we can send 4 bits per symbol. So, the first two bits belong to the first 4ASK signal along with the I axis, and the second two bits belong to the second 4ASK signal along with the Q axis and also encoding the bits into the ASK levels is done using the I branch and Q branch ASK mapping tables as illustrated in Fig. 3.15.

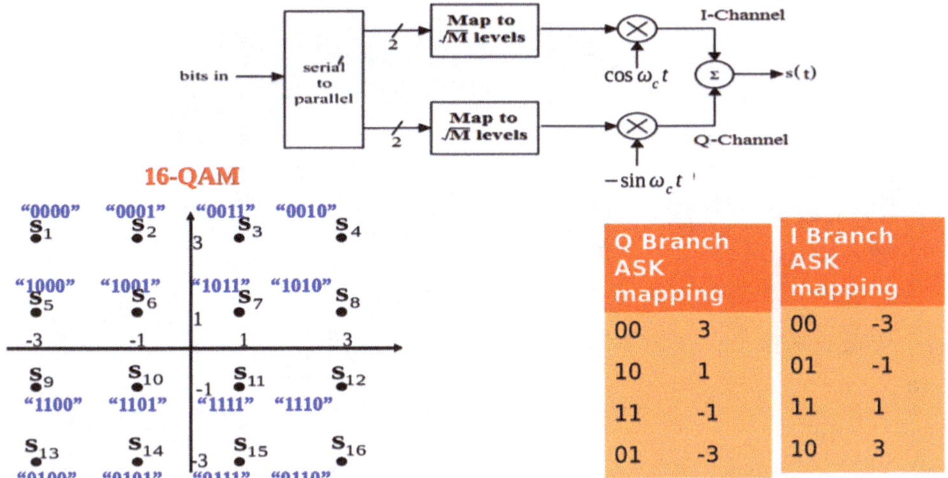

Fig. 3.16 The 16QAM modulation scheme

3.14 The 16QAM Modulation Scheme

In the previous section we observed that the 16QAM signal can easily be generated using two 4ASK modulators. The block diagram is shown in Fig. 3.16 can generate the 16QAM signal.

If at the input we have "1011", the first two bits (11) belong to the I-ASK signal and it would be sent to the I-ASK modulator and then it would be mapped to 4 levels which in this case it would be mapped to the level of 1 and we will module it with the cos(ω_ct) carrier. Similarly, the next two bits (10) would be sent to the Q-ASK modulator and then it would be mapped to 4 levels which in this case it would be mapped to the level of 1 and we will module it with the -sin(ω_ct) carrier. Then we add the I-ASK and Q-ASK signals to get the 16QAM symbol.

3.15 The 16QAM Demodulation

Now, we can easily demodulate the MQAM using two ASK demodulators. The first ASK demodulator will be used to demodulate the I-ASK signal and the second demodulator would be used to demodulate the Q-ASK signal. When the symbol is received in the receiver, it will be multiplied by the cos(ω_ct) carrier and then it would be passed through the LPF and at the output of that LPF we will get the I-level corresponding that received symbol. In the I-axis the dotted lines indicate the thresholds. Since we have the 4ASK signal we will have 3 thresholds and the received symbol will be lied between two thresholds in the I-axis and we can obtain the I-level of the received

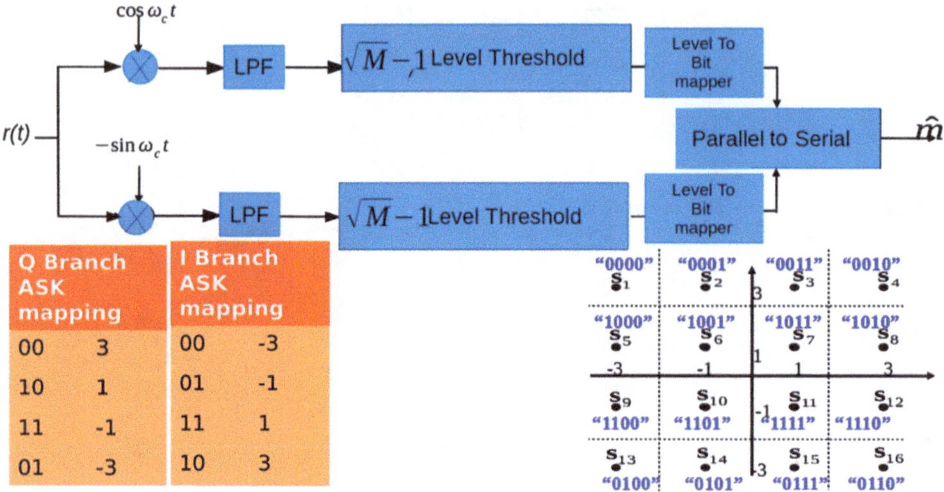

Fig. 3.17 The 16QAM demodulation scheme

symbol. Then from I branch ASK mapping table we can view the result of level to bit mapper. Similarly, we will multiply the received symbol by the -sin(ω_ct) carrier and then it would be passed through the LPF and at the output of that LPF we will get the Q-level corresponding that received symbol. In the Q-axis the dotted lines indicate the thresholds. Since we have the 4ASK signal we will have 3 thresholds and the received symbol will be lied between two thresholds in the Q-axis and we can obtain the Q-level of the received symbol. Then from Q branch ASK mapping table we can view the result of level to bit mapper. Finally, we will pass the outputs of level to bit mapper for I and Q channels to parallel to serial converter and we can get the bits were transmitted by the transmitter as depicted in Fig. 3.17.

3.16 RF Transmitter and Receiver Modules

Now, the 433 MHz RF transmitter and receiver module are available as illustrated in Fig. 3.18.

The RF transmitter and receiver modules are operating in carrier frequency of 433 MHz and they use the ASK modulation scheme. The range of the system is maximum 3 meters without an antenna, but its range can be extendable up to 100 meters by using a small hookup wire as an antenna.

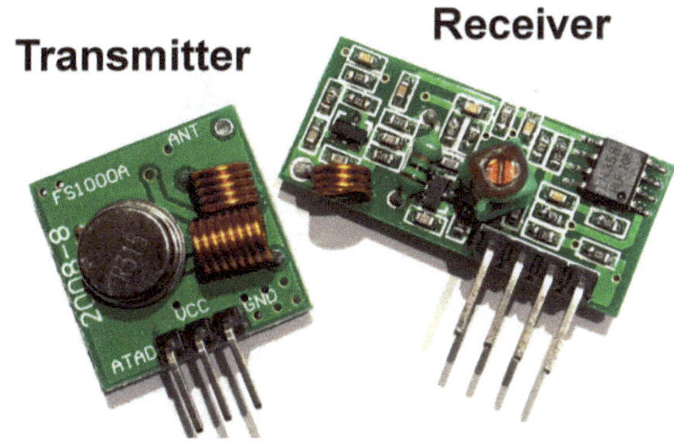

Fig. 3.18 The 433 MHz RF transmitter and receiver modules

3.17 Conclusion

This chapter has provided a comprehensive journey from bandwidth-efficient analog techniques to the foundational digital modulation schemes that form the backbone of contemporary wireless communication. Beginning with Analog Quadrature Amplitude Modulation (QAM), we demonstrated the principle of spectral efficiency by transmitting two independent signals orthogonally within the same bandwidth. This concept serves as a crucial bridge to the digital domain.

The chapter methodically unpacked the core families of digital modulation. We explored how information is encoded in the amplitude with Amplitude Shift Keying (ASK), in frequency with Frequency Shift Keying (FSK)—noting its valuable noise resilience—and in phase with Phase Shift Keying (PSK). Each scheme was analyzed through its generation, constellation representation, and demodulation processes, building a clear framework for understanding how bits are mapped to symbols and recovered in the presence of noise and channel effects.

The culmination of this exploration was the synthesis of these simple techniques into powerful, high-efficiency modulations. We demonstrated that complex schemes like Quadrature Phase Shift Keying (QPSK) and, most importantly, M-ary Quadrature Amplitude Modulation (MQAM) are not entirely novel constructs but are elegantly built by combining in-phase and quadrature channels modulated with BPSK and ASK principles, respectively. This modular perspective demystifies advanced modulations and highlights the scalable nature of digital communication design. The chapter concluded by connecting this theory to tangible hardware, introducing practical RF modules that implement these very schemes.

In summary, Chap. 3 has established the essential vocabulary and operational principles of digital modulation. By understanding ASK, FSK, PSK, and their quadrature combinations, the foundation is now set to delve into the critical performance metrics that govern these systems, such as bit error rate, spectral efficiency, and the impact of noise, which will be the focus of subsequent chapters.

References

1. Tiwana M (2021) RF concepts, Components and Circuits for Beginners. Udemy Inc., San Francisco, CA
2. Couch LW (2013) Digital and analog communication systems, 8th edn. Pearson, Boston

Some Basic Electronics Concepts

4

Contents

4.1 Current Controlled Active Components

The electronic components are divided into 2 main classes, the first one is the active components and the second one is the passive components. To understand the active components, let us examine the transistor that is being used as an amplifier. The transistor has three terminals which are emitter, base and collector. As you can see in Fig. 4.1, the emitter is grounded that is because we would say that transistor is working as a common emitter (CE) amplifier. The collector is connected to the supply V_{CC} through the resistor R_2, and the collector current I_C is flowing through the collector and the base connected to the supply V_{in} through the resistor R_1 and the base current I_B is flowing through the base. Now if we keep the supply V_{CC} constant and vary the current I_B, then we can see that the collector current I_C is also being varied.

So, in this case the collector current I_C is being controlled by the base current, and that is why we call the transistor as an active device because here the transistor is relying on the

M. Pakdel, *Understanding RF Systems*, Synthesis Lectures on RF/Microwaves,
https://doi.org/10.1007/978-3-032-19227-1_4

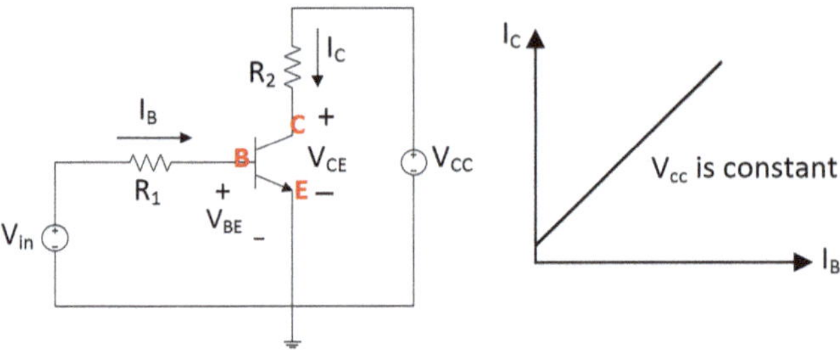

Fig. 4.1 The transistor as a current controlled active component

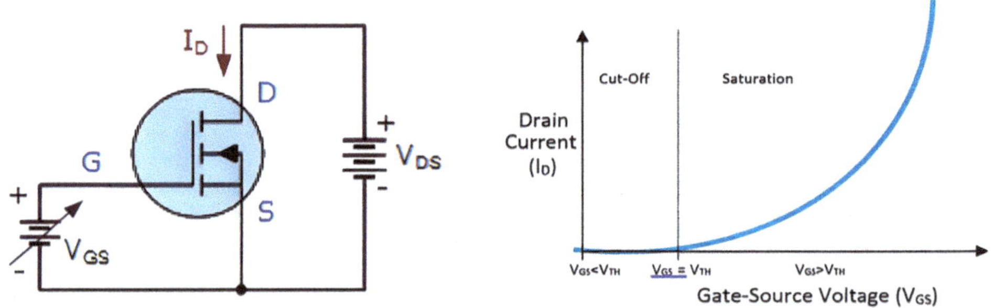

Fig. 4.2 The MOSFET as a voltage controlled active component

base current I_B to control the collector current I_C. So, the active components are those components that rely on an external power source to control or modify electrical signals. In this transistor amplifier circuit, since the collector current is controlled by the base current, we would say that transistor is the current controlled device [1].

4.2 Voltage Controlled Active Components

A MOSFET amplifier is a voltage-controlled device. The MOSFET is a special type of transistor as shown in Fig. 4.2.

The MOSFET has three terminals which are drain (D), gate (G), and source (S) and the source is connected to the ground and that is why we call that MOSFET amplifier is in the common source (CS) configuration. The drain is connected to the voltage supply V_{DS}, and the current is flowing through the drain is I_D. The variable voltage supply V_{GS} is connected to the gate (G) of the MOSFET. Now if the V_{GS} is zero, no current would be flowing and when we increase the V_{GS} still no current flows till it reaches to the certain threshold (V_{TH}) and after that threshold when we increase the V_{GS}, the drain current I_D increases in the nonlinear manner as demonstrated in Fig. 4.2. So, in this case we can see the drain current depends upon the voltage source V_{GS}. Increasing the V_{GS} would cause the increasing of I_D so that is why a MOSFET amplifier is called as a voltage-controlled device and that MOSFET is an active component or an active device [1].

4.3 The Passive Components

Now, to understand the passive components let us consider the example of the circuit which includes a resistor connected to the power supply V and the current flowing through the circuit is I as depicted in Fig. 4.3. The current I flowing in the circuit is determined by the ohm's law (V = IR), or we have I = V/R. If we suppose the voltage supply V be constant then the current I, depends upon the value of resistance R. So, there is not any external source that is determining of how much current would flowing through the circuit as depicted in Fig. 4.3.

However, in the MOSFET example of Sect. 4.2, it was the voltage V_{GS}, that was determining the drain current and in the case of transistor as an amplifier in Sect. 4.1, it was the base current that was determining the collector current. In this case there is not any external source that is determining what current is flowing through the circuit, rather that it depends upon the resistance value when the voltage source is fixed. So, that is why we call that resistor as a passive component or a passive device. The other examples of the passive components are inductors, capacitors, transformers, diodes, etc. [1].

4.4 The Capacitive Reactance

We suppose a capacitor is connected to a voltage source as illustrated in Fig. 4.4.

Now, the capacitor will oppose the current that is flowing to that capacitor and that opposition to the current flow is called the reactance which is defined by the following equation.

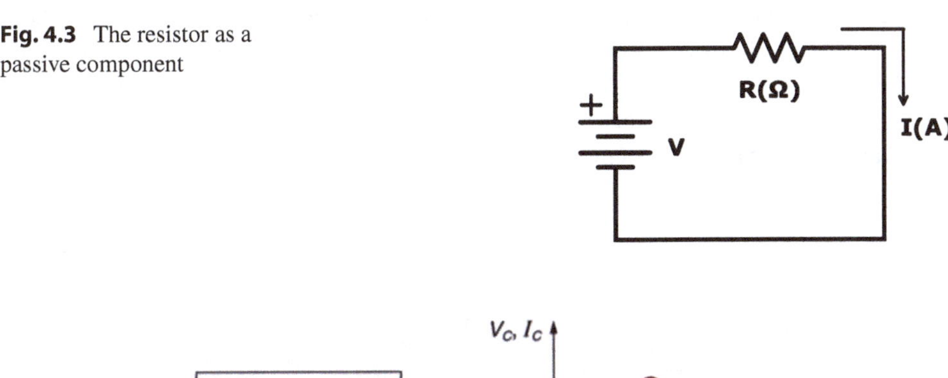

Fig. 4.3 The resistor as a passive component

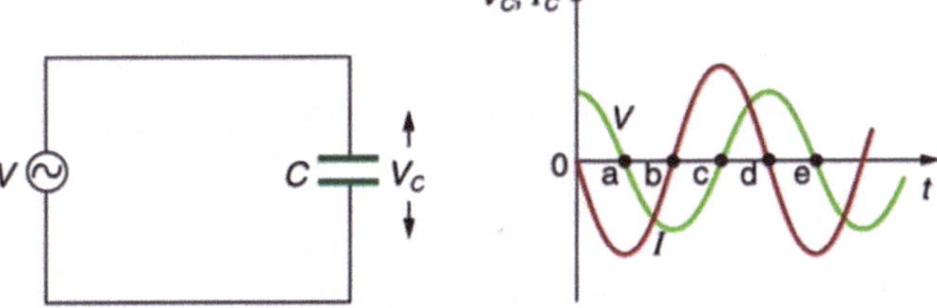

Fig. 4.4 The voltage lag from the current in capacitor

$$X_C = \frac{1}{2\pi fC} \tag{4.1}$$

where f is the frequency of the voltage source. The reactance of capacitor is inversely proportional to the frequency and that means increasing the frequency of the voltage source would cause decreasing of reactance. Also, the capacitor voltage lags the current by 90° phase [2].

4.5 The Inductive Reactance

We suppose an inductor is connected to a voltage source as shown in Fig. 4.5.

Now, the inductor will oppose the current that is flowing to that inductor and that opposition to the current flow is called the reactance which is defined by the following equation.

$$X_L = 2\pi fL \tag{4.2}$$

where f is the frequency of the voltage source. The reactance of inductor is directly proportional to the frequency and that means increasing the frequency of the voltage source would cause increasing of reactance. Also, the inductor current lags the voltage by 90° phase [2].

4.6 The Impedance

Now, in the circuit if you have a resistor, an inductor, or a capacitor or you have both inductor and capacitor, so in those cases the opposition to the current flow called as the impedance and the impedance is the complex number which it has a real part and it is equal to the value of the resistance in the circuit and it has an imaginary part which is equal to the value of reactance in the circuit as given by the following equation.

$$Z = R + jX, \quad X = X_L - X_C \tag{4.3}$$

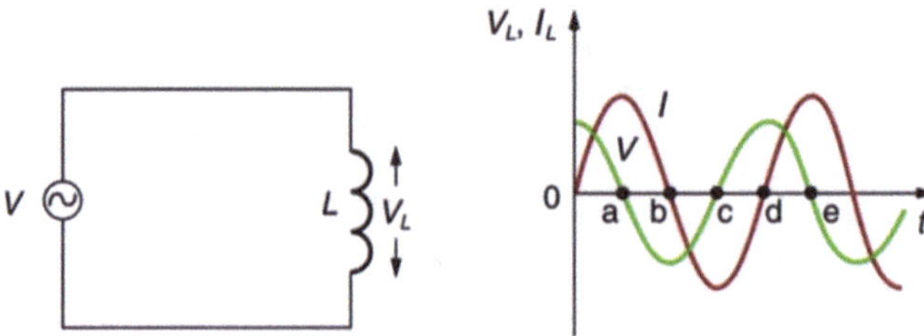

Fig. 4.5 The current lag from the voltage in inductor

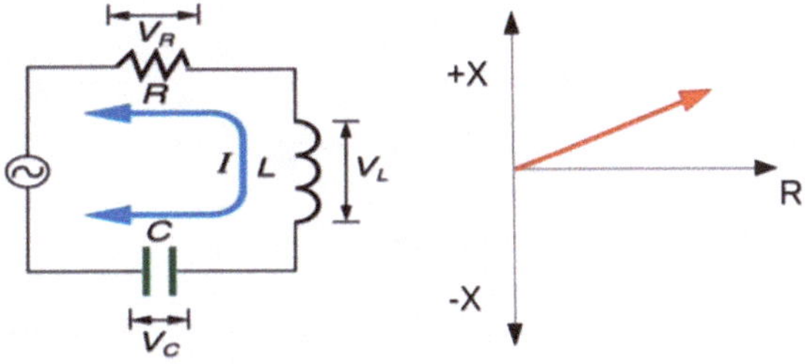

Fig. 4.6 The impedance Z

The impedance Z is complex and has two parts, a real part R which it does not change with frequency and an imaginary part X which it changes with frequency and also the inductive reactance is positive and the capacitive reactance is negative as depicted in Fig. 4.6.

Since the impedance is a complex number, the complex numbers can be represented as vectors, so the impedance also can be represented as a vector and this vector would have two components, one is the imaginary component that would be the reactance status there in the circuit and the second component is the real part which is the resistance status there in the circuit. Now, if the inductive reactance is more in that circuit in that case the vector of impedance would be in the positive direction but if the capacitive reactance is more dominant in that case that vector would be in the negative direction. If you take an example of dummy load which is mostly resistive in nature so in that case the impedance of the dummy load is almost constant with frequency. However, if you take the example of antenna, we know that the antennas are designed for different frequencies, so the impedance of the antenna changes with the frequency [2].

4.7 The Series Resonance in RLC Circuit

Now, in the case of series RLC circuit, the total impedance is given by the following equation.

$$Z = R + j\left(X_L - X_C\right) \tag{4.4}$$

We know that the X_C is inversely proportional to the frequency while X_L is directly proportional to the frequency so at lower frequencies, the X_C is more than X_L, that means our impedance is mainly dominated by the capacitive reactance, while at the higher frequencies, the X_L which is directly proportional to the frequency would exceed the X_C so our impedance becomes dominated by the inductive reactance. However, between these two extremes there is a certain frequency which is called as the resonance frequency at which X_L and X_C become equal and cancel out one another so in that case the impedance is equal to the resistance ($Z = R$), un other words, we can say at that point (f_r), the impedance has the minimum value which is equal to the value of the resistance in the circuit as illustrated in Fig. 4.7.

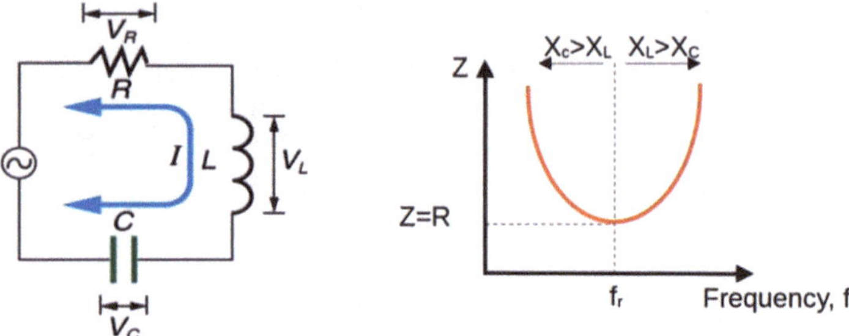

Fig. 4.7 The impedance of series resonance circuit

Now, the resonance frequency f_r can be easily calculated using the following Equation [2].

$$f_r = \frac{1}{2\pi\sqrt{LC}} \tag{4.5}$$

So, the resonance frequency depends upon the inductor L(H) and capacitor C(F).

4.8 The Parallel Resonance in RLC Circuit

Now, in the case of parallel RLC resonance circuit, the resonance frequency is given by the following equation.

$$f_r = \frac{1}{2\pi\sqrt{LC}} \tag{4.6}$$

It is the same formula given in the case of series RLC circuit and in that resonance frequency, $X_L = X_C$ and what happens is that the current is circulating back and forth between the inductor and the capacitor and zero current and energy is being drawn from the supply at the resonance frequency so that parallel LC circuit behaves as an open circuit. So, effectively at the resonance frequency the only part of the impedance that is left in the circuit is that resistance as shown in Fig. 4.8.

So, as you can see in Fig. 4.8, at the resonance frequency the impedance is equal to resistance. However, at the other frequencies, that parallel LC circuit is not behaving as an open circuit. So, in that case the current is flowing in three branches and therefore the impedance is lower. So, we can say that the behavior of the parallel resonance is opposite to the series resonance and in the case of parallel resonance, the maximum impedance is at the resonance frequency while in the case of series resonance the minimum impedance is at the resonance frequency [2].

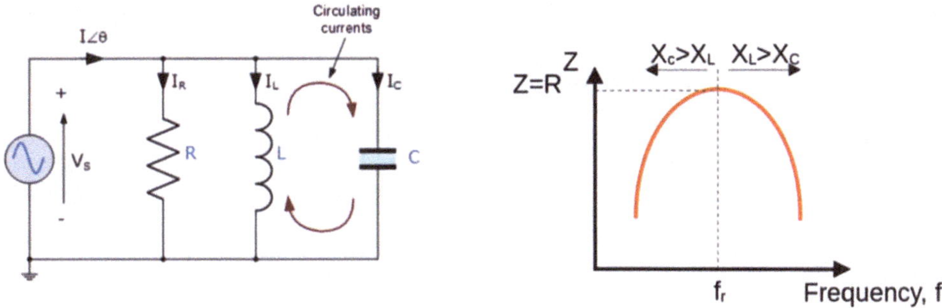

Fig. 4.8 The impedance of parallel resonance circuit

4.9 Conclusion

This chapter has established the fundamental electronic principles that serve as the essential building blocks for all RF circuits and systems. We began by categorizing electronic components into their two foundational classes. We explored active components, such as the BJT and MOSFET, which rely on an external power source to control a signal, with the BJT acting as a current-controlled device and the MOSFET as a voltage-controlled device. In contrast, we examined passive components like resistors, which do not control a signal but respond to it according to their fixed properties.

The chapter then progressed to the critical frequency-dependent behaviors that dominate RF design. We defined reactance, explaining how capacitive reactance (X_C) decreases with frequency while inductive reactance (X_L) increases. Combining resistance with reactance led us to the comprehensive concept of impedance (Z), a complex quantity that fully describes a component's opposition to alternating current and is central to circuit analysis at radio frequencies.

Finally, we applied these concepts to a cornerstone of RF engineering: resonance. By analyzing series and parallel RLC circuits, we demonstrated how at a specific resonant frequency (f_r), inductive and capacitive reactances cancel. This results in two powerful and complementary behaviors: a series resonant circuit exhibits minimum impedance, while a parallel resonant circuit exhibits maximum impedance.

In summary, this chapter has provided the essential vocabulary and analytical tools—understanding active vs. passive control, reactance, impedance, and resonance. These concepts are not merely academic; they are the practical foundation required to design, analyze, and troubleshoot the core RF components that follow, such as amplifiers, oscillators, filters, and impedance matching networks. With this groundwork firmly in place, we are now prepared to explore the specific circuits that bring RF systems to life.

References

1. Tiwana M (2021) RF concepts, components and circuits for beginners. Udemy Inc., San Francisco, CA
2. Nilsson JW, Riedel SA (2015) Electric Circuits, 10th edn. Pearson, Boston

The Oscillators

5

Contents

5.1 The Function of an Oscillator

The oscillator is used in order to generate the carrier frequency and that carrier frequency is in the sinusoidal form as shown in Fig. 5.1.

The oscillator is an important part of RF circuits because it is used to generate the carrier frequency and the carrier frequency is necessary for modulation in the transmitter and demodulation in the receiver [1].

M. Pakdel, *Understanding RF Systems*, Synthesis Lectures on RF/Microwaves,
https://doi.org/10.1007/978-3-032-19227-1_5

Fig. 5.1 The oscillator in RF transmitter

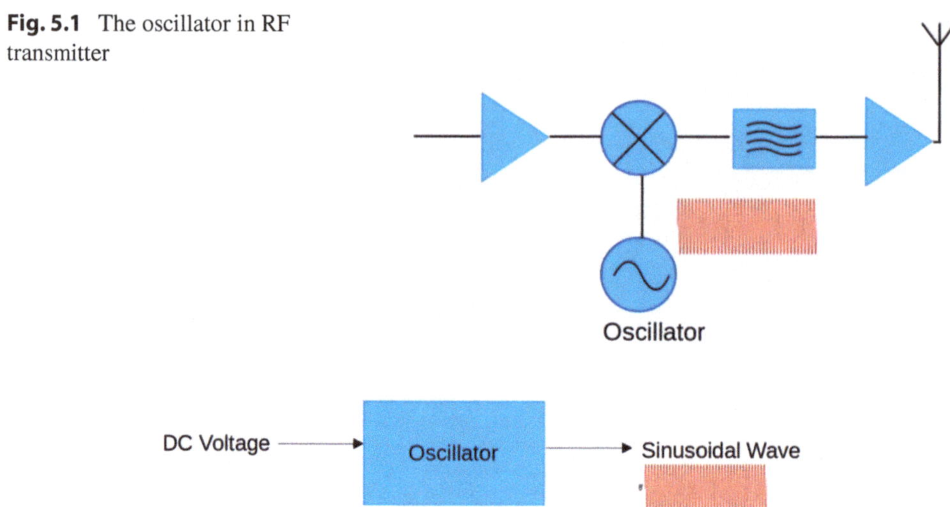

Fig. 5.2 The linear (harmonic) oscillator scheme

5.2 Types of Oscillators

The oscillators can be classified into two categories, the first one is the linear (harmonic) oscillators and the second one is the nonlinear (relaxation) oscillators. The output of linear oscillators is a sinusoidal waveform so these are the oscillators that are used in the RF circuits, while the output of nonlinear oscillators is non sinusoidal, so these oscillators are mostly used in the digital circuits. Since we are interested in RF circuits in this book so we are only going to discuss the different types of linear oscillators and we would not discuss about nonlinear oscillators [2].

5.3 Types of Linear (Harmonic) Oscillators

In the linear oscillators, the DC voltage is as their input and at the output they produce a sinusoidal waveform and these linear oscillators are further subdivided into two main types the first one is the feedback oscillators and the second one is the negative resistance oscillators as depicted in Fig. 5.2.

In the next sections, we would discuss about these types of oscillators.

5.4 The Feedback Oscillators

In this section we would discuss about the feedback oscillators and they are called feedback oscillators because they employ the positive feedback circuit. The general block diagram of the positive feedback circuit is illustrated in Fig. 5.3.

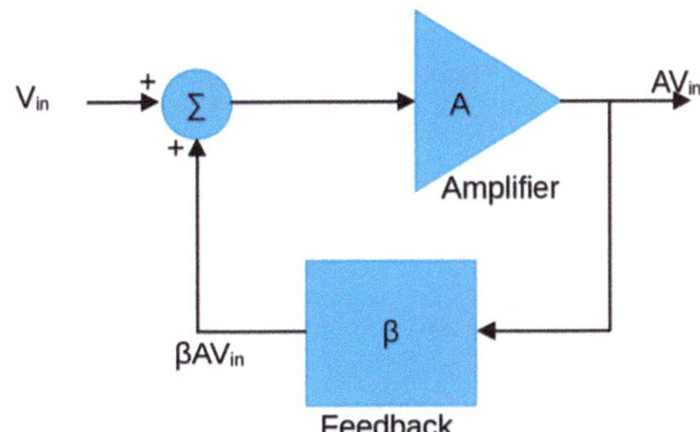

Fig. 5.3 The general block diagram of the positive feedback circuit

The first component of it is called the adder, the second component is an amplifier and whatever input to this amplifier, the amplifier amplifies it by the factor of A and obviously A is greater than 1. Now, a part of amplifier output is fed back to the adder through the feedback block and what part of the amplifier output is fed back to the adder is determined by the factor β and obviously β < 1. Now, suppose we apply a sinusoidal waveform in the V_{in} and since we have initially zero output at the output of the feedback, it would go to the input of the amplifier and the amplifier would amplify it and we would get AV_{in} at the output of the amplifier that means the V_{in} has been amplified by the factor of A. Now, the output of AV_{in} is fed back to the feedback block and at the output of the feedback block we would have βAV_{in}, and the βA is called the loop gain and the loop gain may be smaller than 1, equal to 1 or greater than 1depending on the value of β and A. Now, if the overall loop gain of this feedback loop is 0o and βA is greater than 1, in that case the feedback that would be appearing in the adder input would be greater than the V_{in}. Now, at the same time if we remove the V_{in}, then that feedback would again go to the amplifier and pass through the feedback and come back again to the adder and it would be further amplified and then that amplified feedback would go to the amplifier and pass through the feedback block and appear to the input of the adder and it would be further amplified again as shown in Fig. 5.4.

As you can see in Fig. 5.4, the initial amplitude gives sine and it is increasing until the amplifier reaches its saturation point and at the saturation point since the amplifier can provide no more gain after the saturation, so we get the distortions and we are not getting the pure sinusoidal waveform that we want. The second condition is where the βA < 1, in that case the feedback that we are getting at the input of the adder would be smaller than Vin and if we remove the V_{in}, that feedback would go to the amplifier and then it would pass to the feedback circuit and since βA < 1, that feedback would be further reduced and then it would pass through the loop back circuit and it would be further reduced again. So, we can see the amplitude of the sine wave that we are getting there at the output is decreasing. So, in both cases when βA > 1 or βA < 1, and even if the overall phase shift of the circuit is 0o, we are not getting the sustain oscillations. The conditions for the sustain oscillations are called as the Barkhausen criteria and the first condition is that the loop gain

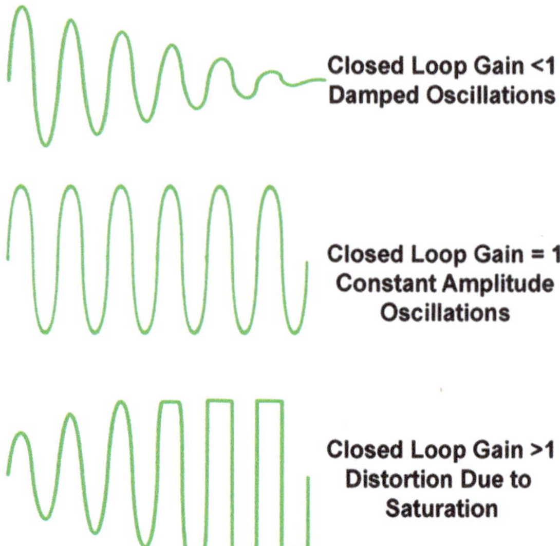

Fig. 5.4 The oscillations due to the value of βA

βA should be equal to 1 (βA = 1) and the second condition is that the feedback should be in phase with the V_{in}. So, if those conditions are met in that case when we apply the V_{in} at the start, we would get feedback there in the input of the adder and it would have the same amplitude as V_{in} and it would have the same phase, now if we remove the V_{in} in that case, we would get the same feedback value there and at the output we would be getting the sinusoidal waveform that has the same amplitude. So, in that condition we would have the sustained oscillations and we are generating a sustain sinusoidal wave at the output. Now, in the practical feedback oscillator circuits there are losses so for that purpose the value of βA is not exactly equal to 1 rather it is slightly higher than 1 in order to compensate those losses. Also, the practical feedback oscillator circuits produce the sustain oscillations even if no V_{in} is applied and it is possible due to the white noise in the components of the feedback circuit, because we know that the white noise has all the frequencies and for one particular frequency it would meet the Barkhausen criteria and that means the overall phase shift of the feedback circuit would be 0o and the βA = 1 rather it is slightly higher than 1. So, for that particular frequency, there would be a sustain oscillations at the output of the feedback oscillator circuit and we would get our desired sinusoidal waveform at that particular frequency that meets the Barkhausen criteria.

5.5 Types of Feedback Oscillators

The feedback oscillators are further subdivided into three types, the first one is the RC oscillators, and those are really low frequency oscillators that start from few kilo hertz and go till few megahertz. Then you have the crystal oscillators which start from few kilo hertz and they can go up till hundred megahertz and also you have the LC oscillators that can go up to a few hundred of megahertz and now we are going to discuss about these oscillator types in more details.

5.6 RC Phase Shift Oscillators

Now, in the case of RC phase shift oscillator for the amplifier we normally use a transistor or an operational amplifier (op-amp) and for the feedback circuit we use a three stage RC circuit as depicted in Fig. 5.5.

Since, we use the RC circuit in the feedback, that is why we call it as the RC phase shift oscillator. An example of the RC phase shift oscillator that is using an op-amp as the amplifier as illustrated in Fig. 5.6.

The positive input of the amplifier is grounded while the negative input of the amplifier is being used and the amplifier is working in the inverting configuration, that means between the input and the output there is a phase difference of 180°. Now, according to the Barkhausen criteria, the overall phase shift of the feedback circuit must be 0°. So, the RC phase shift section should provide the phase of 180° and that means each of the RCs should provide the phase shift of 60°, so that the overall phase shift would be 180° for total

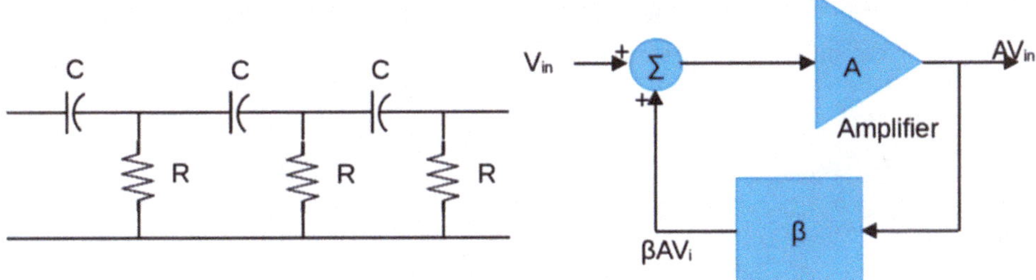

Fig. 5.5 The RC phase shift oscillator scheme

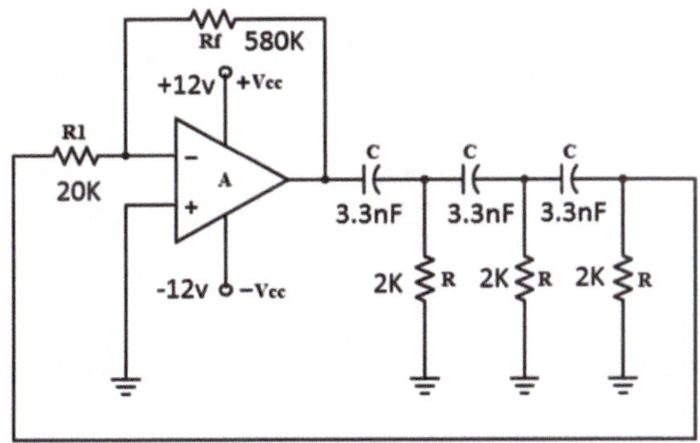

Fig. 5.6 An example of the RC phase shift oscillator

RC stages. Then we combine that phase shift with the phase shift of the amplifier so that the overall phase shift of feedback circuit becomes 0° and the first requirement of Barkhausen criteria is met. The second requirement of Barkhausen criteria is that if the value of β for this RC circuit is 1/29 then for the op-amp by adjusting the resistor R_f which is the feedback resistor and the resistor R1, the gain of the amplifier should be adjusted in such a way that A = 29. So, that overall gain or the gain of the feedback loop gain must be equal to 1 (Aβ = 1) and the circuit would oscillate at the frequency for which the Barkhausen criteria is met and that oscillation frequency is calculated by the following equation.

$$f_{osc} = \frac{1}{2\pi RC\sqrt{6}} \tag{5.1}$$

So, that means that oscillation frequency depends upon the values of R and C and at the output we would get a sinusoidal waveform that has that oscillation frequency. We can also use a transistor as an amplifier, as shown in Fig. 5.7.

The transistor is working in the common emitter (CE) configuration and in the common emitter configuration between the input (base) and the output (collector), there is a phase difference of 180°, so we know that according to the Barkhausen criteria the overall phase shift should be 360° or 0°. So, the remaining 180° phase shift is provided by the RC stages such that each RC stage provides 60° phase shift. So, in this way from the output at the collector and then going back to the RC feedback circuit to the input (base), the overall phase shift is 0° and the resistors R_4 and R_5 are adjusted such a way that the overall gain of that feedback circuit which is βA be equal to 1 (βA = 1).

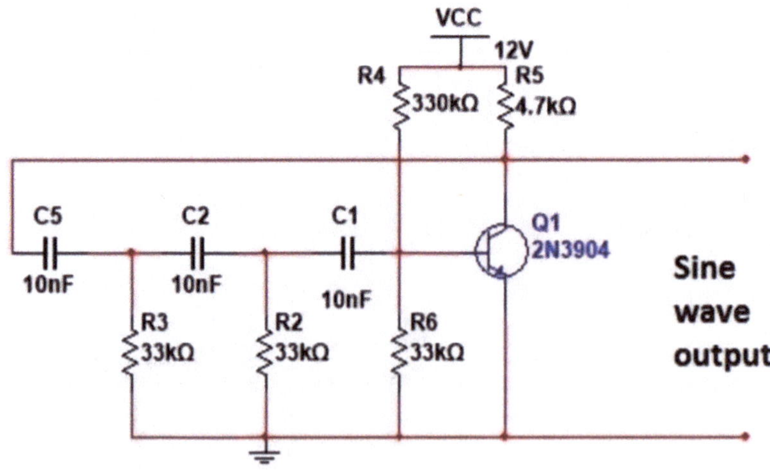

Fig. 5.7 The RC phase shift oscillator using a transistor as the amplifier

5.7 The LC Tank Circuit

Now, in this section we discus about the LC tank circuit in LC oscillators. The LC oscillators use the LC tank circuit. An LC tank circuit is an inductor that is connected in parallel to a capacitor. So, in order to understand the LC oscillators, we need to understand the operation of the LC tank circuit and initially that capacitor is connected to the voltage source V and it would completely charge the capacitor as depicted in Fig. 5.8.

Now, once the capacitor is fully charged, we change the switch position from A to B and the capacitor would start discharging and the current would start flowing in that direction. So, as the capacitor is discharging, the energy that stored in that capacitor would start transferring to that inductor and it would be stored in the inductor in the form of magnetic field, also the current is decreasing by discharging of the capacitor and we know that that inductor opposes the decreasing in the current through it and the back EMF or voltage would appear which would have the inverse polarity of voltage source V. Once, the capacitor is completely discharged and now all the energy that was there in the capacitor would be stored in the inductor. So, the inductor would serve now as a voltage source and it would start charging the capacitor and the direction of the current flow would be the same. So, the inductor charging the capacitor and once the capacitor is fully charged the polarity of the capacitor will be inverted and all the energy that was there in the inductor is now transferred to the capacitor. Now, the capacitor is starting to discharge but the flow of the current would be in the opposite direction and the energy of capacitor is being transferred to the inductor and it is stored in the inductor in the form of magnetic field and once that capacitor is completely discharged that inductor would serve as a voltage source with the same polarity of voltage source V and it would start charging the capacitor until the capacitor is fully charged and after that the whole procedure would repeat again, and the capacitor would discharge and the current would flow in the opposite direction and the energy would be stored in the inductor and the inductor would charge that capacitor and so on. So,

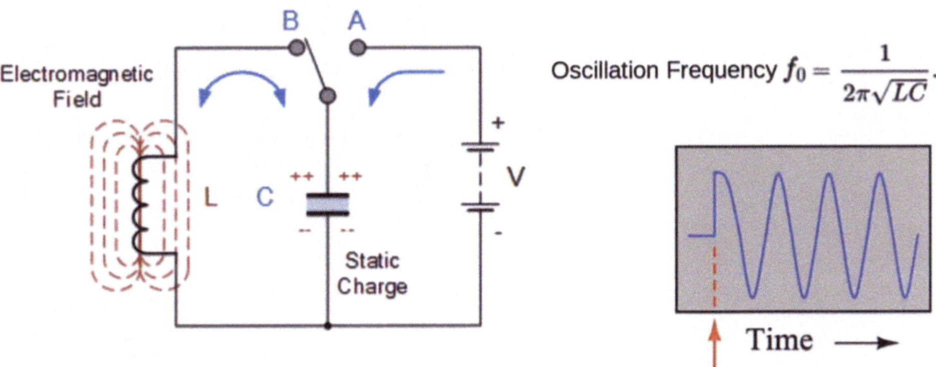

Fig. 5.8 The LC tank circuit

we can see that the oscillations are produced in the circuit and that oscillation would have the particular frequency which is given by the following equation.

$$f_{osc} = \frac{1}{2\pi\sqrt{LC}} \tag{5.2}$$

That means that oscillation depends upon the values of L and C. However, we know that in the practical circuits the ideal capacitor and ideal inductor do not exist and they will have some resistance in them and as a result as the current flow to the resistors the energy is dissipated in the resistance so the amplitude of those oscillations would start decreasing and ultimately the oscillations would die out. So, for the sustained oscillations, we need to have some external source that is compensating for that dissipation of energy.

5.8 The LC Hartley Oscillator

Now, in this section we will discuss about the Hartley oscillator which is a type of LC oscillators. A transistor-based Hartley oscillator circuit is illustrated in Fig 5.9. So, it is using a transistor as an amplifier, and that amplifier is used to provide the energy to the LC tank circuit so, that oscillations would not die out in the LC tank circuit.

The transistor in the common emitter (CE) configuration that means there is between input (base) and output (collector) a phase shift of 180°, so the remaining 180° phase shift is provided by that LC tank circuit, and in the LC tank circuit the inductor is a coil that is the center tapped, and that means the inductor can be divided into two coils that are close to one another and the center tap is grounded. So, the 90° phase shift is provided by the first coil and the other 90° phase shift is provided that second coil. So, the total phase shift

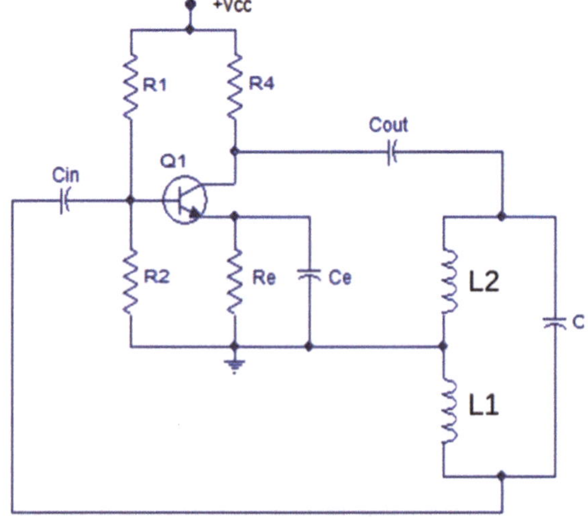

Fig. 5.9 A transistor-based Hartley oscillator circuit

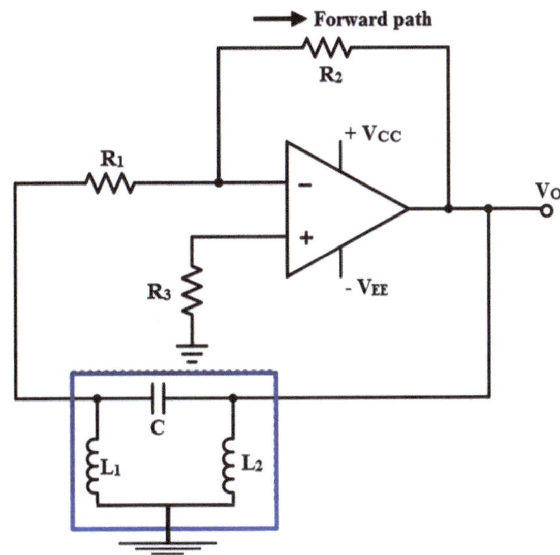

Fig. 5.10 The op-amp based Hartley oscillator

of the feedback circuit can be calculated as 180° phase shift that is provided by the transistor and 90° phase shift that is provided by L2 coil and the other 90° phase shift that is provided by the L1 coil, so the total phase shift is 360° and that is the first requirement of Barkhausen criteria for sustained oscillations. The second requirement is $\beta A = 1$, and we have $\beta = L1/L2$. So, we can adjust the feedback factor β by changing the center tap and once the Barkhausen criteria is met then the oscillations would be produced in the circuit at the output and the frequency for which that Barkhausen criteria is met and the oscillations would be produced, it is given by the Eq. (5.2) and that means it depends upon the total inductance ($L = L1 + L2$) in the tank circuit and the capacitor that there is in the LC tank circuit. Now, another Hartley oscillator configuration that is using the op-amp as the amplifier is shown in Fig. 5.10.

As you can see in Fig. 5.10, the positive terminal of that amplifier is grounded and the negative terminal is being used so that op-amp is being used in the inverting configuration that means that between the input and the output there is a phase shift of 180° and the remaining 180° phase shift is provided by that LC tank circuit so that the total phase shift is 360° or 0° and the amount of feedback is $\beta = L1/L2$ and we adjust the value of feedback factor β such that $\beta A = 1$.

5.9 The LC Colpitts Oscillator

Now, the Colpitts oscillator is another type of LC oscillator that is very similar to the Hartley oscillator except that the capacitor in the Hartley oscillator is replaced by the inductor in the Colpitts oscillator and the coils L1 and L2 in the Hartley oscillator are replaced by the capacitors C1 and C2 in the Colpitts oscillator as depicted in Fig. 5.11, and

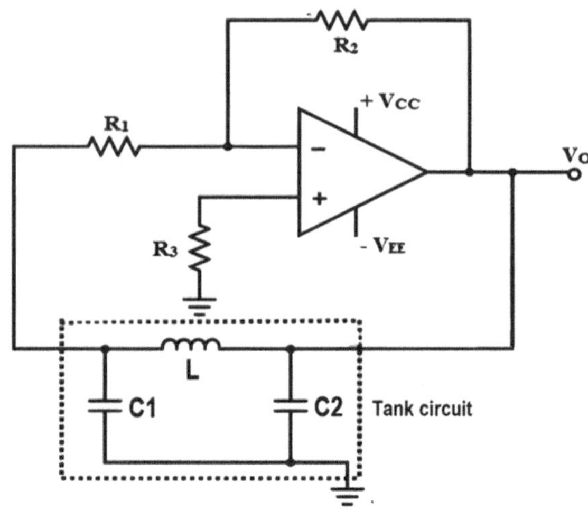

Fig. 5.11 The op-amp based Colpitts oscillator

using the capacitors C1 and C2 instead of coils L1 and L2 is easier because it is much easier to tune those capacitors instead of tuning the center tap of those coils.

For the Colpitts oscillator, we have the following equations:

$$\beta = \frac{C2}{C1} \tag{5.3}$$

$$f_{osc} = \frac{1}{2\pi\sqrt{LC}}, \quad \frac{1}{C} = \frac{1}{C_1} + \frac{1}{C_2} \tag{5.4}$$

5.10 The Crystal Oscillators

There are problems with RC and LC oscillators as we operate with RC and LC oscillators and the temperature changes the value of R, L and C also change and as a result the oscillation frequency changes and similarly when over of period of time in which we use those oscillators with the passage of time the value of R, L and C change and as a result their oscillation frequency changes and those oscillators need to be re-tuned, on the other hand the crystal oscillators are more stable and their oscillation frequency is not changed with the temperature and when they are operated in the period of time their resonant frequency also does not change much. So, another advantage of these crystal oscillators is that they have a high Q factor in the range of 10,000 to 20,000 or they are more frequency selective. In order to understand the Q factor, let us consider the diagram in Fig. 5.12.

The f_r is the oscillation frequency of an oscillator and that oscillator would resonate at that frequency with maximum power, however the oscillator will oscillate in other frequencies that near to the resonance frequency but with a lesser power. Now, if the Δf is the

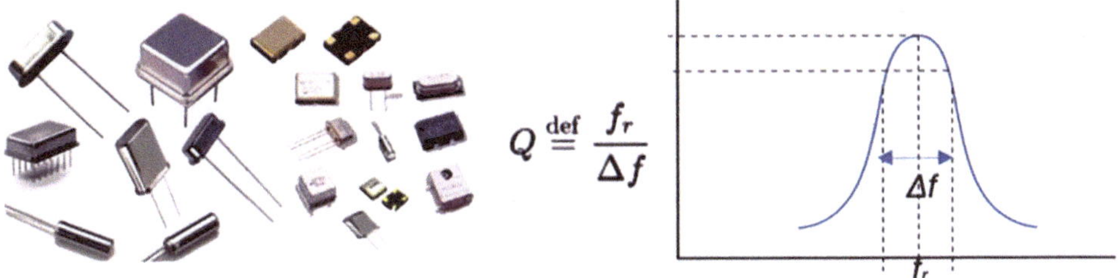

Fig. 5.12 The crystal oscillators characteristics

Fig. 5.13 The piezoelectric effect

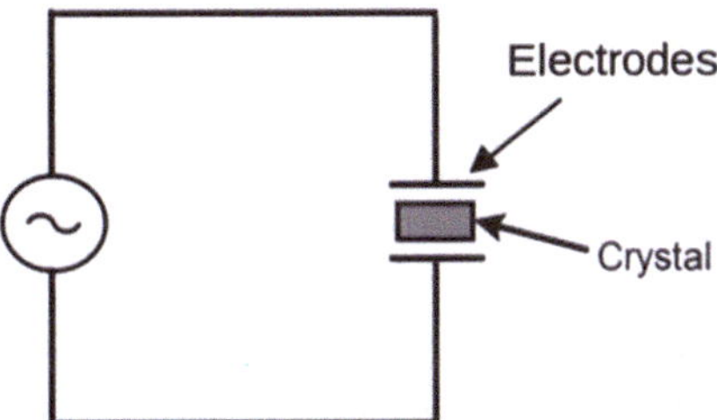

bandwidth between the two points where the oscillator provides oscillations but with half power as compared to the maximum power and we have the following equation.

$$Q \stackrel{\text{def}}{=} \frac{f_r}{\Delta f} \tag{5.5}$$

If the curve in Fig. 5.12 is narrower, in that case the Q factor is higher and that means the oscillator is more frequency selective and that gives us a more precise resonance frequency. Also, those crystal oscillators come in the shapes of Fig. 5.12 and they can provide us the oscillation frequencies from 40 kHz to up till 100 MHz depending upon at which frequency we tune those oscillators.

5.11 The Piezoelectric Effect

The operation of the crystal oscillator is based upon the piezoelectric effect and this effect is observed when the crystal is made of quartz, tourmaline, or Rochelle salt. As you can see in Fig. 5.13, there are two electrodes which are connected to the crystal and if we apply an AC current to that crystal and that AC current has a certain frequency, that crystal would also start to vibrating in that frequency so that is called the piezoelectric effect. There is also the inverse piezoelectric effect, and we suppose that the crystal is vibrating with a certain frequency, then it would produce the current that has the same frequency as the vibrating frequency of that crystal.

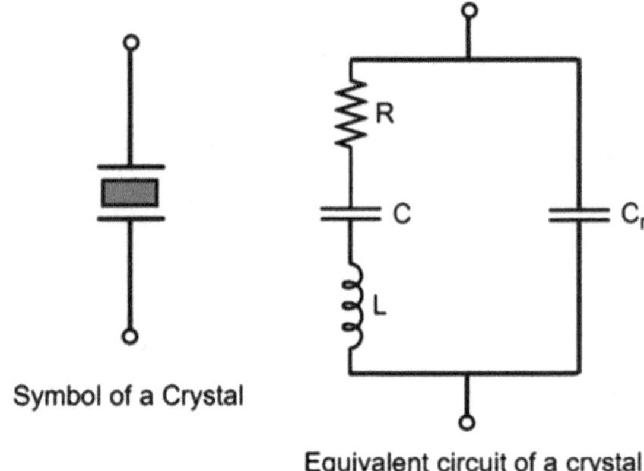

Fig. 5.14 The electrical equivalent of quartz crystal

5.12 The Electrical Equivalent of Quartz Crystal

The electrical equivalent of quartz crystal is illustrated in Fig. 5.14.

The C_m is the shunt capacitance and that depends upon the connection of electrodes to the quartz crystal. The R is the equivalent resistance of that quartz crystal and then you have the C which is the motional capacitance and L is the motional inductance and those parameters are due to the vibration of the quartz crystal and they are depends upon the elasticity of the quartz crystal, the area of the plates, and the thickness of the crystal.

5.13 The Crystal Oscillator Circuits

In the equivalent circuit of quartz crystal in Fig. 5.14 there are a series RLC circuit and a parallel RLC circuit in it so the quartz crystal can oscillate at the series resonant frequency that we are denoting as fr1, and at the parallel resonant frequency that we are denoting as fr2 and we know that in the case of series resonance the impedance of series RLC circuit will be minimum and in the case of parallel resonance, the impedance of the circuit would be maximum as shown in Fig. 5.15.

Now, the crystal oscillator not only can oscillate in f_{r1} and f_{r2} but it can also oscillate at the multiple harmonics of these frequencies, for example $2f_{r1}$, $3f_{r1}$ and so on, and also $2f_{r2}$, $3f_{r2}$ and so on. Now, if we want to operate the crystal in the series resonance operation, then we can use the type of the circuit as depicted in Fig. 5.16a, we use the transistor as an amplifier and it is operating in common emitter configuration (CE) that means between the input (base) and the output (collector) there is a phase shift of 180° and between the output and the input there is a feedback path, and in this feedback path we have placed the crystal so the remaining 180° phase shift is provided by this crystal, so the total phase shift of this feedback circuit is 360° or 0°. Now, if we want to operate tat crystal in the parallel resonance operation, then we can use the type of circuit in Fig. 5.16b and we can very well see

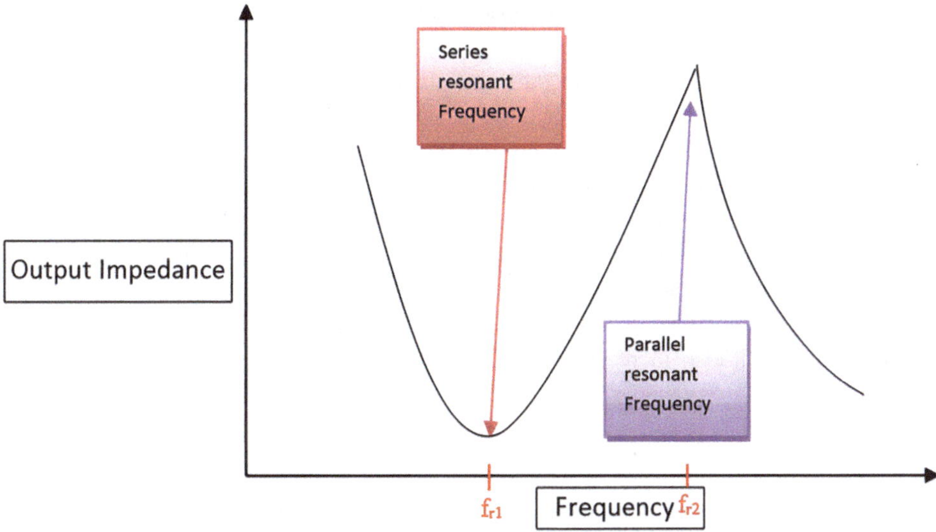

Fig. 5.15 The series and parallel resonance of the crystal oscillator

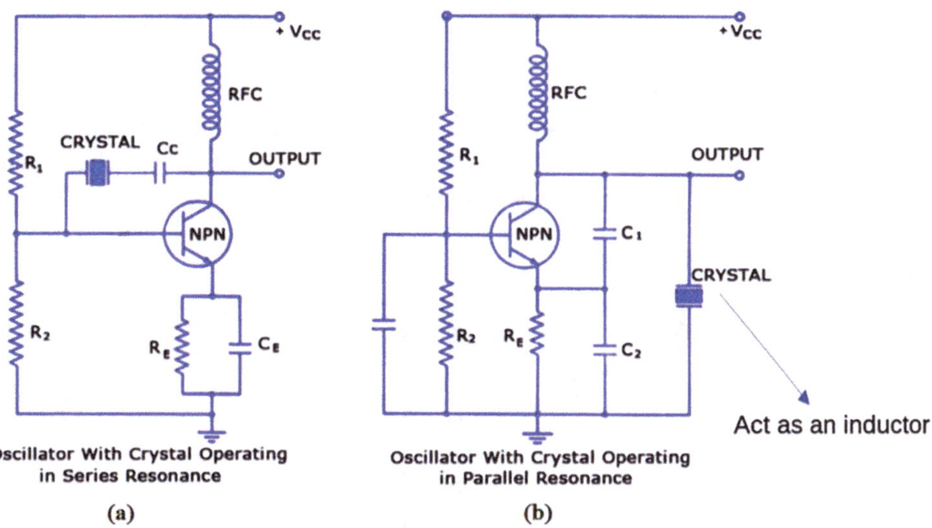

Fig. 5.16 The use of cristal oscillator in series and parall resonance

that this circuit is very similar to the circuit of Colpitts oscillator. In the Colpitts oscillator we use an LC tank circuit in which we have two capacitors and we have an inductor and there instead of inductor we are using a crystal and that crystal would be acting as an inductor and it would be providing 180° phase shift in the feedback and the other one 180° phase shift is provided by the transistor, so the total phase shift is 360° or 0° in the feedback circuit and the output is the collector.

5.14 The Negative Resistance Oscillator

The tunnel diode can be used as a negative resistance oscillator and that tunnel diode has the negative resistance region, because if the V_D be the voltage across the tunnel diode, and between the V_D values of V_P and V_V there is a region that is called the negative resistance region and, in this region, if the value of V_D across the diode is increased then in that case the resistance of the diode is also increased and that is why it is called the negative resistance region. In order to know how the tunnel diode is producing oscillations we need to know the circuit as illustrated in Fig. 5.17.

In Fig. 5.17, when the switch is closed the current I_D through the circuit increases and the voltage drop V_D also increases through the tunnel diode, so we have the following equation.

$$V_D = I_D R_D \tag{5.6}$$

At the point A, the current has its peak value since if V_D is further increased in that case the tunnel diode enters a negative resistance region and as a result the resistance of a diode increases as you can view in Eq. (5.6). Since the source voltage V is divided between tunnel diode and resistance R, so the R_D increases and the V_D also increases so as the V_D increasing you can see in Fig. 5.17 that the current I_D is decreasing until we reach to the value of V_V, So, when $V_D = V_V$, the tunnel diode leaves the negative resistance region and that means further increasing in V_D, the R_D decreases, the voltage V_D decreases as compared to the voltage drop through resistance R and V_D is pushed back to the negative resistance region and after that we start increasing the V_D again and it leaves the negative resistance region and then it is pushed back to the negative resistance region, so in this way

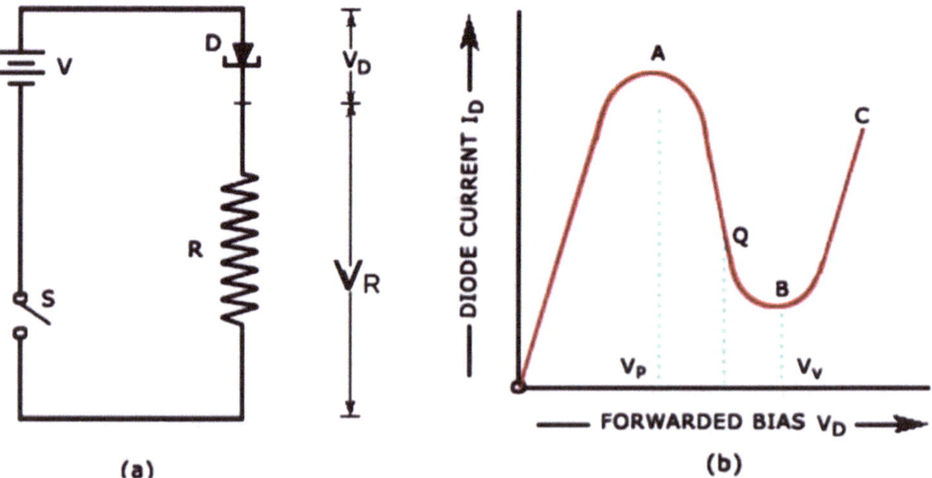

Fig. 5.17 The tunnel diode characteristics

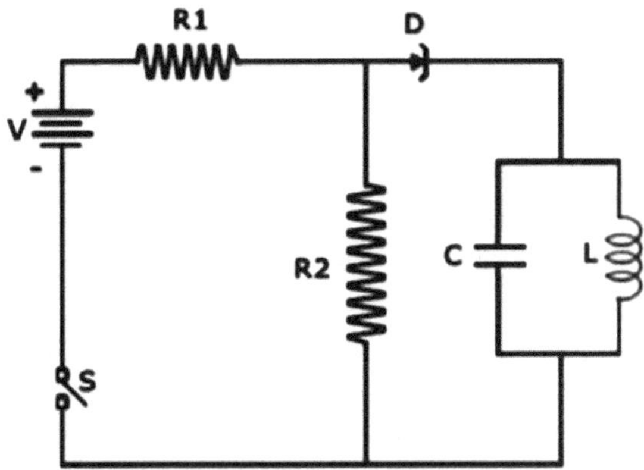

Fig. 5.18 Another tunnel diode based negative resistance oscillator

the oscillation take place around point B in Fig. 5.17. So, the V_D moves back and forth between negative and positive region and producing oscillations. We have also the series resistance R and the voltage V is divided between V_D and VR, so we can adjust the resistance R in order to set the bias point or operating point of that tunnel diode at point Q and that tunnel diode would produce oscillations around that operating point. So, the value of R, at the bias point is calculated by the following equation.

$$R = \frac{V - V_{DQ}}{I_{DQ}} \tag{5.7}$$

Now, we have another tunnel diode based negative resistance oscillator as shown in Fig. 5.18.

As you can see in Fig. 5.18, we have an LC tank circuit that is connected in series with the tunnel diode and there R1 and R2 are used to set the bias point of that tunnel diode and that LC tank circuit would resonate at its resonance frequency and it would produce oscillations at the resonance frequency so that tunnel diode based circuit would produce oscillations at the resonance frequency of that LC tank circuit and those oscillators are used to generate really high frequency oscillations ranging from 0.5 GHz to 40 GHz.

5.15 The Voltage Controlled Oscillator (VCO)

In the case of the voltage-controlled oscillator, the oscillation frequency is controlled by the input voltage. The block diagram of VCO is depicted in Fig. 5.19.

The equation of VCO in a very simple way is as below.

$$\omega = \omega_c + v_{in}(t) \tag{5.8}$$

Fig. 5.19 The block diagram of voltage-controlled oscillator (VCO)

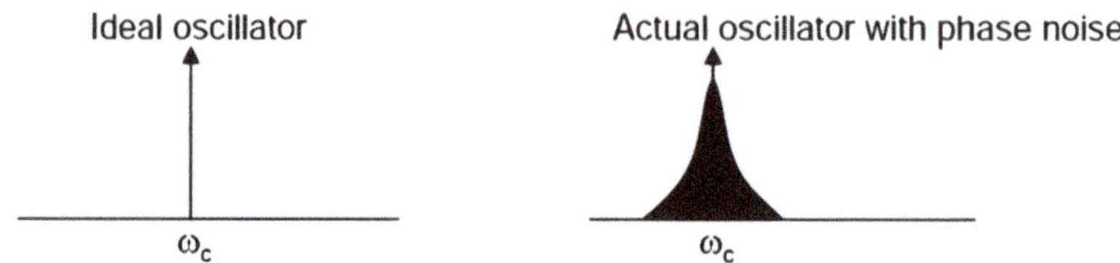

Fig. 5.20 The frequency response of an ideal oscillator and actual oscillator with phase noise

The v_{in} is the voltage that is being applied to the VCO and ω is the output frequency of VCO, now if v_{in} is 0, in that case the output frequency of VCO (ω), would be equal to the carrier frequency (ω_c). Now, if the v_{in} is increasing the ω is also increasing and as v_{in} is decreasing the ω is also decreasing, so the output frequency of VCO depends upon the input voltage and that is why we call it the voltage-controlled oscillator, and one of the important applications of VCO is in the generation of frequency modulated (FM) signal because in that case the change in the amplitude of the signal is modulated as the change of the frequency which is precisely what the VCO is doing.

5.16 The Problem of Phase Noise in Oscillators

The phase noise caused by random fluctuations in the phase of an oscillator waveform, and we have the following equation.

$$\theta = \omega t \tag{5.9}$$

That means when the θ has fluctuations the frequency ω also fluctuates. The frequency response of an ideal oscillator and actual oscillator with phase noise is illustrated in Fig. 5.20.

As you can see in Fig. 5.20, the ideal oscillator has no phase noise and there are no fluctuations in its frequency, but in the case of oscillator with phase noise we can see its frequency is fluctuating around ω_c. Now, we suppose the modulated signal (wanted signal) that we want to demodulate it and near that desired signal there is an interfering signal (interferer) and if we use the ideal oscillator to demodulate the desired signal, we can see that we bring the wanted signal to the frequency of 0 and the interferer would be near to

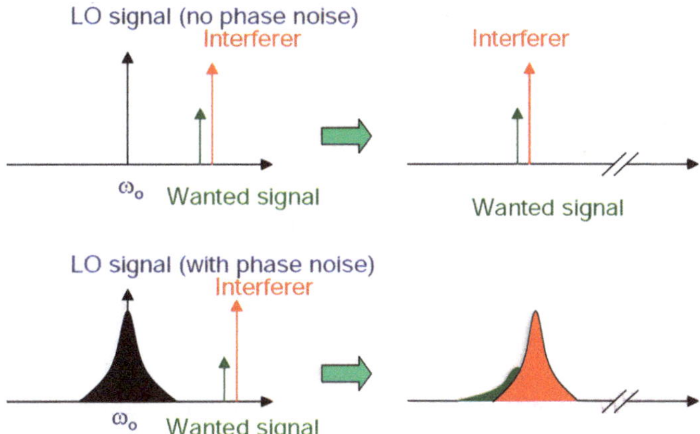

Fig. 5.21 Demodulation using ideal and actual with phase noise oscillators

Fig. 5.22 The phase noise problem on the transmitter side

that wanted signal but we can easily filter out the wanted signal using a filter. However, if we are using an oscillator that has the phase noise and we use that oscillator in order to demodulate the wanted signal then we are bringing the wanted signal to the frequency of 0 but it has fluctuations in its frequency and the interferer has also fluctuations in its frequency and the tail of interferer is interfering with the green signal and we cannot easily filter out the green signal as shown in Fig. 5.21.

The phase noise also cause problems on the transmitter side for example we have a transmitter that is transmitting its signal at the carrier frequency of ω_2 and we have another transmitter that is transmitting its signal at the carrier frequency of ω_1 but that transmitter has noise phase in its oscillator that is used to modulate the signal to ω_1 and as a result that modulated signal would also have fluctuations in its frequency and as a result the tail of the signal would interfere with the signal that is at the frequency of ω_2 as depicted in Fig. 5.22.

5.17 Conclusion

This chapter has provided a systematic and comprehensive study of oscillators, establishing their indispensable role as the origin of the stable frequency references upon which all RF systems depend. Beginning with a foundational classification, we focused on linear harmonic oscillators, detailing the universal Barkhausen criteria that govern the precise conditions for sustained sinusoidal oscillation: unity loop gain and zero-phase shift.

The exploration then progressed through the practical implementation of these principles across a spectrum of technologies. We analyzed RC phase-shift oscillators for lower frequencies, transitioned to LC oscillators like the Hartley and Colpitts configurations for higher-frequency applications, and culminated with the superior stability of crystal oscillators, whose high Q-factor and temperature insensitivity make them the cornerstone of precision timing. The chapter also examined specialized architectures, including negative resistance oscillators for microwave frequencies and the tunable Voltage-Controlled Oscillator (VCO) for modulation applications.

Crucially, the chapter concluded by addressing a key non-ideal characteristic that limits real-world system performance: phase noise. We demonstrated how phase noise degrades both receiver selectivity and transmitter spectral purity, highlighting that the quality of an oscillator is defined not only by its frequency of oscillation but also by the spectral cleanliness of its output.

In summary, this chapter has equipped you with the principles to understand, select, and analyze the various oscillator types that generate and control the carrier signals in RF circuits. This knowledge forms a critical foundation for the subsequent chapters on mixers, amplifiers, and phase-locked loops, where the stable signal generated here is manipulated, amplified, and synchronized to enable complex RF communication.

References

1. Tiwana M (2021) RF concepts, components and circuits for beginners. Udemy Inc., San Francisco, CA
2. Steer MD (2019) Microwave and RF design: amplifiers and oscillators. NC State University, Raleigh

The Mixer and Its Applications

6

Contents

6.1 The Mixer Operation

The mixer is used to change the frequency of the RF signal. For example, at the transmitter the mixer is used to change the center frequency of the message signal from 0 Hz to the carrier frequency ω_c and on the receiver side the mixer is used to change the frequency of the modulated signal from the center frequency of ω_c to 0 Hz as shown in Fig. 6.1 [1].

Now, the mixer has two input ports and one output port as depicted in Fig. 6.2.

We are inputting on the first port of the mixer a sinusoidal that has the frequency of ω_1 and in the frequency domain this signal can be shown as an impulse that is located at the frequency of ω_1, and on another input we are inputting a sinusoidal that has the frequency of ω_2, now at the output we get the product of those two signals so at the output we get two signals, the first signal is located at the frequency of $\omega_1 + \omega_2$ and we also get the second signal which is located at the frequency of $\omega_1-\omega_2$. Now, only one of the two output signals is selected using the filter as illustrated in Fig. 6.3.

For example, we can use the upper frequency filter in order to select the signal that is located at the frequency of $\omega_1 + \omega_2$ and since that signal is located at the higher frequency and if we select that signal that procedure is called the up-conversion, and if we use the filter to select the signal that is located at the frequency of $\omega_1-\omega_2$ which is the lower

M. Pakdel, *Understanding RF Systems*, Synthesis Lectures on RF/Microwaves,
https://doi.org/10.1007/978-3-032-19227-1_6

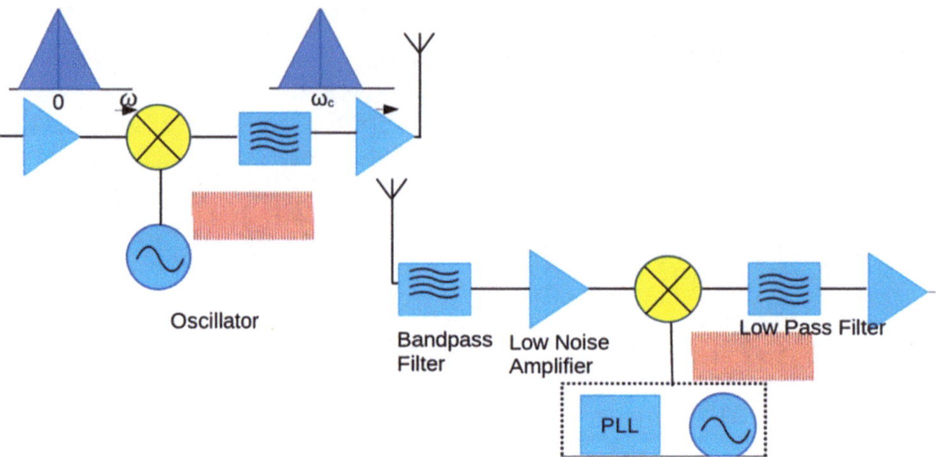

Fig. 6.1 The use of mixer in transmitter and receiver

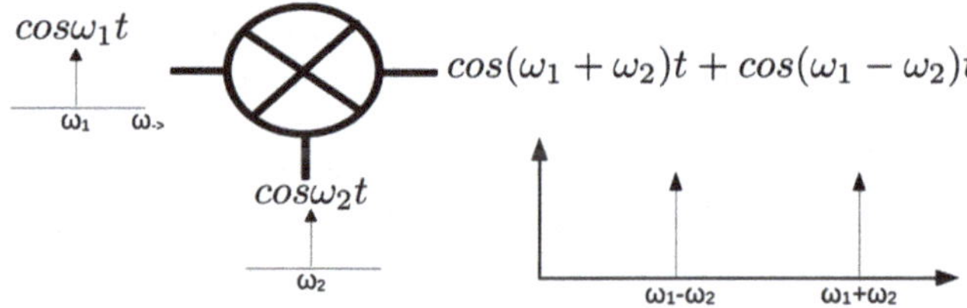

Fig. 6.2 The mixer input and output ports

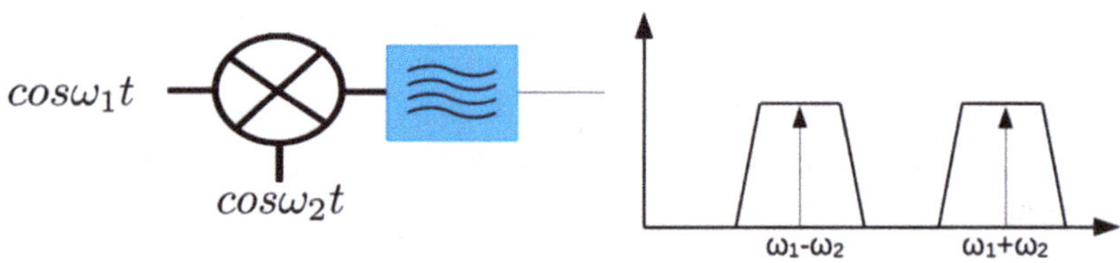

Fig. 6.3 Selecting one output using the filter

frequency in that case this procedure is called as the down-conversion. Now, we have a numerical example of the mixer as shown in Fig 6.4.

In the first input of the mixer, we are inputting a sinusoidal that has the frequency of 500 MHz and at the other input of the mixer we are inputting a sinusoidal that has the frequency of 400 MHz, so at the output, we will get two signals the first signal has the frequency which is difference between of those two input signals and it is equal to 100 MHz and for the second signal the we get it would be equal to the sum of the two input signal frequencies and it is equal to 900 MHz.

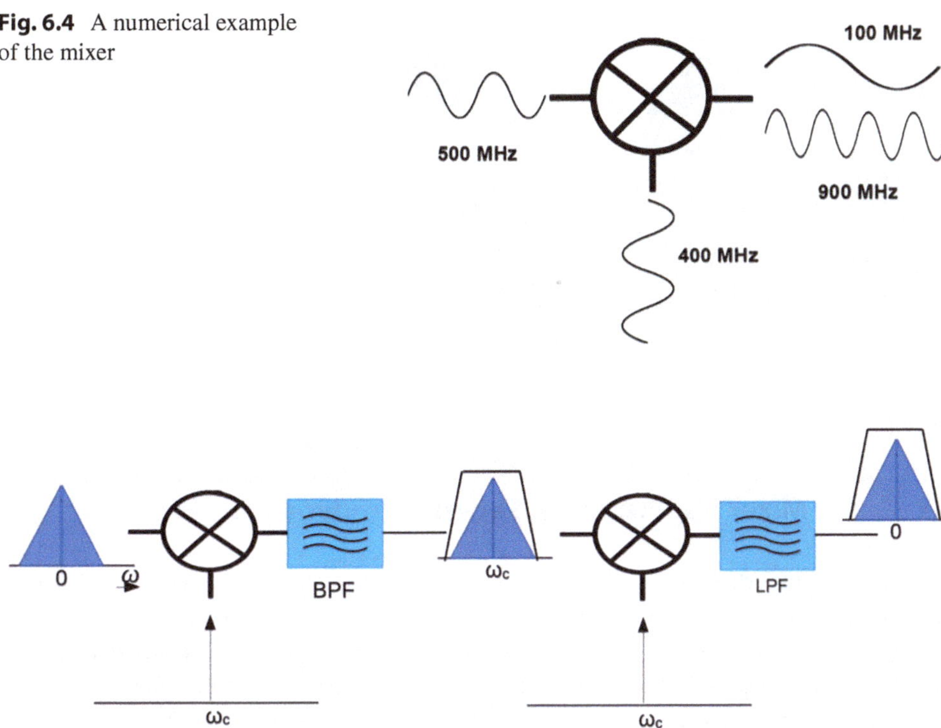

Fig. 6.4 A numerical example of the mixer

Fig. 6.5 The mixer application in Modulation/Demodulation

6.2 The Mixer Application in Modulation/Demodulation

One of the most important applications of the mixer are in the modulation and demodulation procedures. In the modulation at the transmitter, we have the message signal which is located or centered at the frequency of 0 Hz and it is being multiplied by the carrier which is a cosine that is located at the carrier frequency of ω_c, now at the output of the mixer we would get the signal that is located at the frequency of sum of 0 Hz and ω_c, and so we will filter out the signal using the bandpass filter (BPF) which is centered at ω_c and that would be our modulated signal. Now at the receiver, we would again multiply that received signal with the carrier frequency which is a cosine located at the frequency of ω_c and at the output we would get two signals, one signal would be located at the carrier frequency of $2\omega_c$ and the other signal would be located at the difference of the two carrier signals ($\omega_c - \omega_c = 0$ Hz) which is 0 Hz and that is our desired signal which is located or centered at 0 Hz and we will filter it out using the lowpass filter (LPF) as depicted in Fig. 6.5.

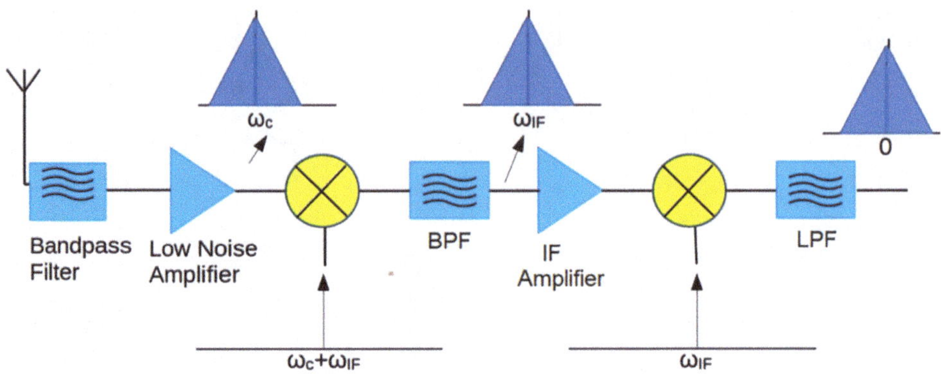

Fig. 6.6 The superheterodyne receiver

6.3 The Mixer Application in Superhererodyne Receiver

Now, another important application of the mixer is in the superheterodyne receiver and in that receiver architecture, the modulated signal is not directly demodulated to the frequency of 0 Hz rather that modulated signal is first multiplied with a carrier that has the frequency of $\omega_c + \omega_{IF}$, where the frequency ω_{IF} is some intermediate frequency. Now at the output of the mixer we get two signals, one of them has the frequency of $2\omega_c + \omega_{IF}$ and the other frequency is ω_{IF} or it is located at the intermediate frequency and we select that signal using a bandpass filter and that signal is then amplified using the IF amplifier and after that the message signal which is at the frequency of ω_{IF}, it is again multiplied with the carrier frequency of ω_{IF} using the mixer and at the output we get two signals, one of them is located at the frequency of $2\omega_{IF}$ and the other is located at 0 Hz, so we select the signal at 0 Hz using a LPF as illustrated in Fig. 6.6 [2].

Now, what is the reason that we are doing that demodulation procedure in two stages? We do that because the low noise amplifier (LNA) is normally amplifying a very high frequency signal and at the higher frequencies the gain is smaller so we cannot use that LNA for providing high gain to its input signal, so what we do is we first change the frequency of the message signal to the intermediate frequency (IF) and we use an IF amplifier that is optimized to provide high gain at the IF, so after amplifying the signal using the IF amplifier then we demodulate the signal to the center frequency of 0 Hz using the second stage mixer.

6.4 Conclusion

This chapter has established the mixer as a cornerstone component in RF systems, enabling the essential function of frequency translation. Beginning with its fundamental operation as a signal multiplier, we explored how a mixer generates sum and difference frequencies from its two inputs, a process that forms the basis for both up-conversion in transmitters and down-conversion in receivers.

The practical significance of this principle was then examined through its direct application in straightforward modulation and demodulation blocks, demonstrating the mixer's role in shifting signals between baseband and passband. The chapter culminated in a detailed analysis of the mixer's pivotal function within the superheterodyne receiver architecture. Here, the strategic use of an Intermediate Frequency (IF) stage was shown to be a critical engineering solution. By employing a mixer to first down-convert the incoming RF signal to a lower, fixed IF, the design enables the use of stable, high-gain IF amplifiers and sharp filters, thereby overcoming the significant challenges of achieving high performance, selectivity, and gain directly at very high radio frequencies.

In summary, this chapter has illustrated that the mixer is far more than a simple multiplier; it is the enabling device for efficient spectrum management and high-performance receiver design. The principles covered here—from basic frequency translation to the architecture of the superheterodyne receiver—provide the necessary foundation for understanding more advanced RF front-end components, frequency synthesizers, and modern receiver topologies discussed in subsequent chapters.

References

1. Tiwana M (2021) RF concepts, components and circuits for beginners. Udemy Inc., San Francisco, CA
2. Losee F (2005) RF systems, components, and circuits handbook, 2nd edn. Artech House, Washington, D.C.

The RF Filters

7

Contents

7.1 Introduction to RF Filters

The RF filters can be represented using the symbols as shown in Fig. 7.1.

The RF filters are used to eliminate signals at unwanted frequencies. For example, we use the filter as depicted in Fig. 7.2 in order to only allow the modulated signal to pass through and to block all other frequencies.

Also in the receiver, we use the bandpass filter (BPF) in order to allow only desired modulated signal to pass through and to block all other frequencies and after the mixer, we are using the lowpass filter (LPF) in order to allow demodulated message signal to pass through and to block all other frequencies as illustrated in Fig. 7.3 [1].

M. Pakdel, *Understanding RF Systems*, Synthesis Lectures on RF/Microwaves,
https://doi.org/10.1007/978-3-032-19227-1_7

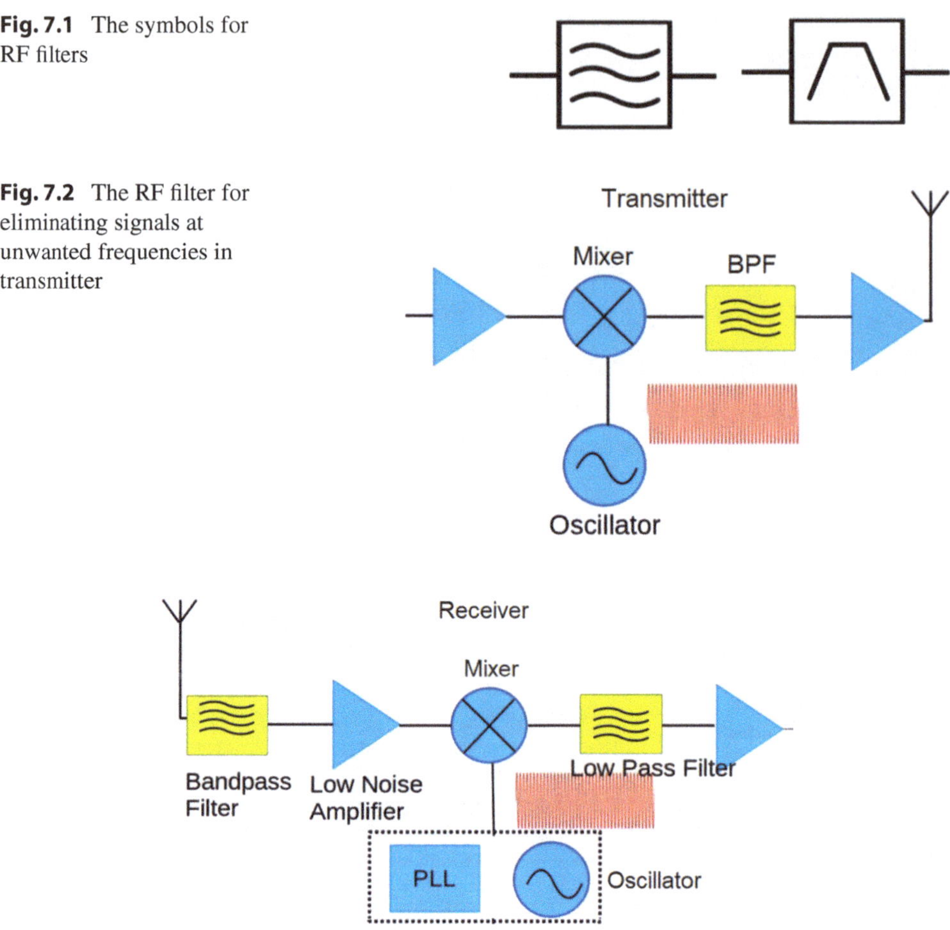

Fig. 7.1 The symbols for RF filters

Fig. 7.2 The RF filter for eliminating signals at unwanted frequencies in transmitter

Fig. 7.3 The RF filter for eliminating signals at unwanted frequencies in receiver

7.2 The RF Filters Classes and Types

There are two main classes of filters, the first one is the passive filters that are made of passive components like capacitors, resistors, inductors, and then you have active filters that they are involving an amplifier, also there are different types of filters you have the lowpass filter (LPF), the frequency response of LPF shows which frequencies that filter is allowing to pass through and which frequencies it is blocking so in the case of LPF we can see that it is allowing low frequencies to pass through while it is blocking the high frequencies and then you have the high pass filter (HPF), that allows the high frequencies to pass through while it blocks the low frequencies and then you have the bandpass filter (BPF) that allows a range of frequencies to pass through while it blocks all other frequencies and

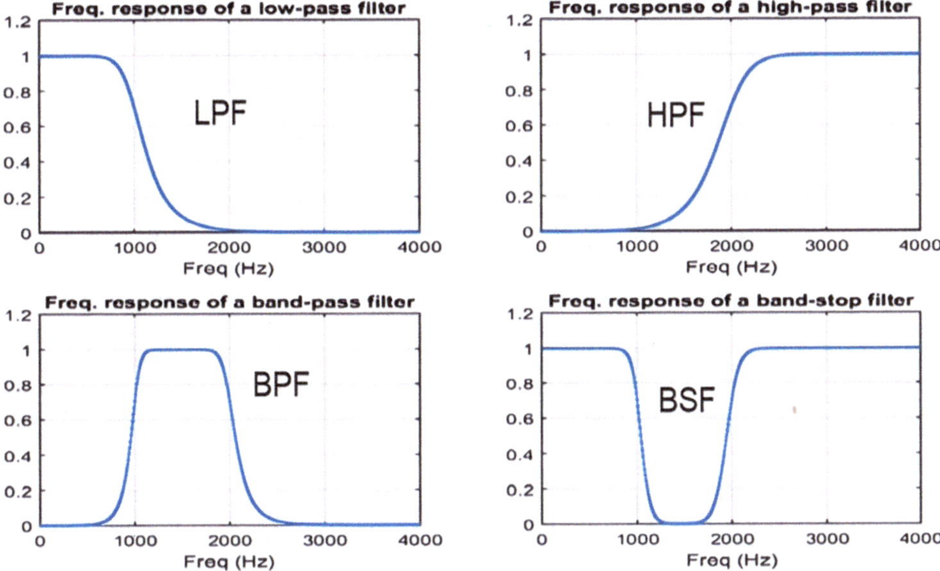

Fig. 7.4 The types of filters

then you have the band stop (BSF) that stops a band or a range of frequencies while it allows the other frequencies to pass through it as shown in Fig. 7.4.

7.3 Measuring Power in RF Systems (dBm)

Now, the RF power is expressed in decibel milli watt (dBm) and we have the following equation.

$$P(dBm) = 10\log_{10}\left(\frac{P(mW)}{1mW}\right) \tag{7.1}$$

So, if we have the power in milli watt we can easily convert it into power in dBm using the Eq. (7.1). So, for example the power of 1 MW in dBm is calculated as the equation below.

$$P(dBm) = 10\log_{10}\left(\frac{10^6 \times 10^3 mW}{1mW}\right) = 90 \tag{7.2}$$

The values of dBm for different power levels are given in Table 7.1.

As you can see in Table 7.1, between 1 MW and 10^{-12} W (10^{-9} mW) in terms of watts there are a huge difference in power in watts but you can see in terms of dBm this variation is only from +90 dBm to −90 dBm, so in terms of numbers that variation in dBm is much more manageable as compared to the power levels in watts. Now, we suppose a mobile

Table 7.1 The values of dBm for different power levels

dBm	mW	Power level
+90	1,000,000,000	1 MW
+80	100,000,000	100 kW
+70	10,000,000	10 kW
+60	1,000,000	1 kW
+50	100,000	100 W
+40	10,000	10 W
+30	1000	1 W
+20	100	0.1 W
+10	10	0.01 W
0	1	0.001 W = 1 mW
−10	0.1	0.1 mW
−20	0.01	0.01 mW
−30	0.001	0.001 mW
−40	0.0001	0.0001 mW
−50	0.00001	0.00001 mW
−60	0.000001	0.000001 mW
−70	0.0000001	0.0000001 mW
−80	0.00000001	0.00000001 mW
−90	0.000000001	0.00000001 mW

Fig. 7.5 The gain definition for a device with input power P_{in} and output power P_{out}

station is transmitting the power as 1 W and the power is received in the base station is 1 mW, so we can see that there is a difference of 1000 times between the transmitting and receiving power and if we use the dBm to express the powers so in that case the difference between 1 W and 1 mW is 20 dBm and 0 dBm and so that is much more manageable.

7.4 Measuring Amplification and Attenuation (dB)

Now, the gain or attenuation of a device is measured in decibel (dB). For example, in Fig. 7.5, we have a device with the input power of P_{in} and the output power of P_{out}, and we can obtain the gain of that device in dB by the following equation.

$$Gain(dB) = 10\log_{10}\left(\frac{P_{out}}{P_{in}}\right) \tag{7.3}$$

Now, if the device is amplifier, then the output power P_{out} is more than the input power P_{in} so would get a positive gain from Eq. (7.3) which is expressed in dBs, on the other hand if that device is for example a filter that is attenuating a frequency, then for that frequency that gain would be in negative dBs by using Eq. (7.3). Similarly, suppose a mobile station is transmitting a signal to the base station and the transmitted signal suffers the path loss that means the signal is attenuated and its power is reduced when it reaches the base station as depicted in Fig. 7.6. So, if we divide the power that is reaching the base station by the transmitted power and calculate the gain in dBs then that gain would be in negative dBs or another words that signal is suffering from attenuation.

7.5 The 3 dB Bandwidth of a Filter

Now, the bandwidth of a filter is often defined as the 3 dB bandwidth. In order to understand the 3 dB bandwidth we need to understand the 3 dB cutoff point. Now, we consider a lowpass filter as illustrated in Fig. 7.7.

At low frequencies the gain of lowpass filter is 0 dB or in the other words we can say that filter is allowing low frequencies to pass through as it is and after that we can see that the gain of the filter is decreasing and that means the filter is now attenuating those

Fig. 7.6 The power loss in the signal transmission (path loss)

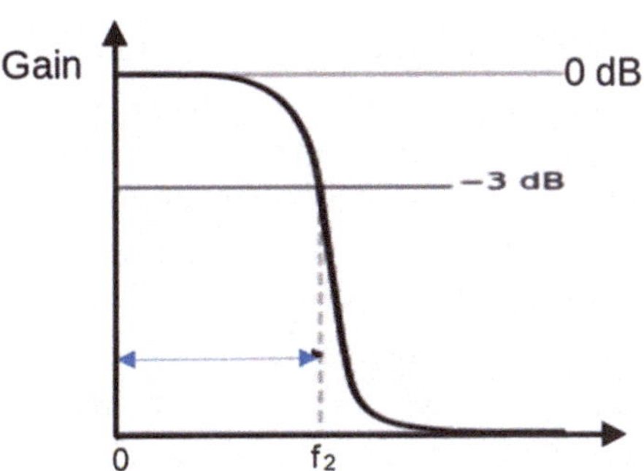

Fig. 7.7 The lowpass filter frequency response

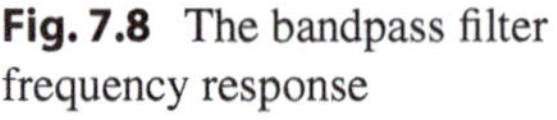

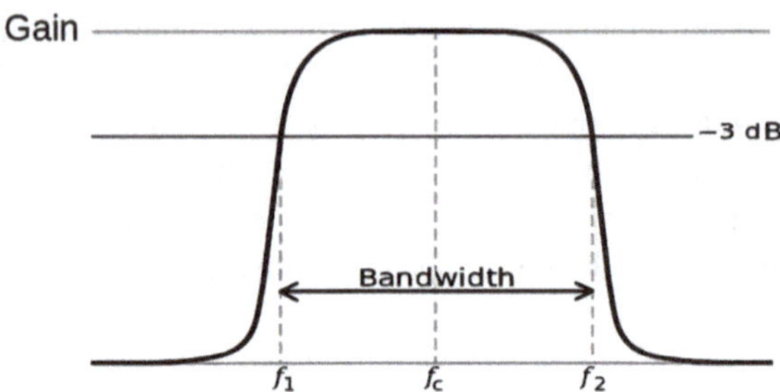

Fig. 7.8 The bandpass filter frequency response

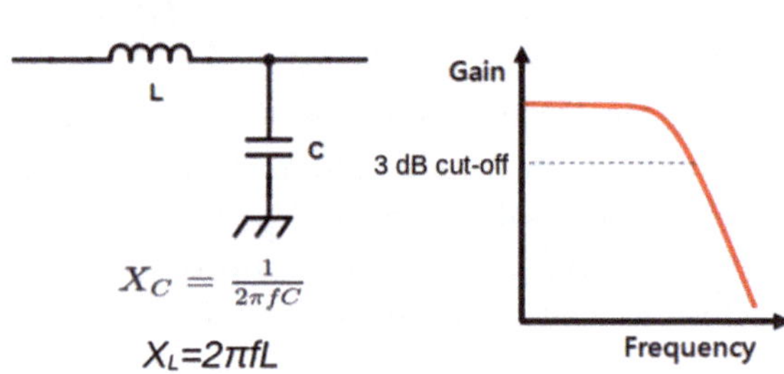

Fig. 7.9 The configuration of LC lowpass filter

frequencies and at a point the gain of the filter is -3 dB or on the other words that frequency has been attenuated to its half power or in the other words that frequency has been reduced to its half power and the bandwidth of a lowpass filter is often measured from 0 to that −3 dB point which is called the 3 dB cutoff point, while in the case of bandpass filters, the bandwidth is measured between two frequency points where the gain of the filter has reduced to −3 dB and the signal passing through the filter is reduced to half its power as shown in Fig. 7.8 [2].

7.6 The LC Lowpass Filter

Now, the first type of passive filters we are going to discuss is the LC lowpass filter and that lowpass filter allows the DC and low frequency signals to pass through and it blocks the high frequency signals and one of the configurations of that LC lowpass filter is depicted in Fig. 7.9.

We know that the reactance of inductor is directly proportional to the frequency and its inductance value and that means we can adjust the inductance in such a way that high frequencies we want to block and for those frequencies that inductor has a very high reactance, or in other words that inductor blocks those higher frequencies and for capacitor we know that its reactance is inversely proportional to the frequency and the capacitance value so we can adjust the capacitance in such a way that the higher frequencies that we want to block for those higher frequencies that capacitor has a very low reactance or in other words that capacitor serves like a short circuit for those higher frequencies and those higher frequencies are grounded so those higher frequencies are not only blocked by that inductor

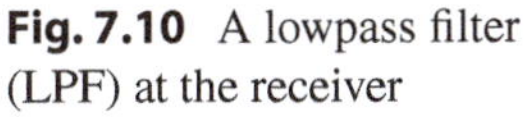

Fig. 7.10 A lowpass filter (LPF) at the receiver

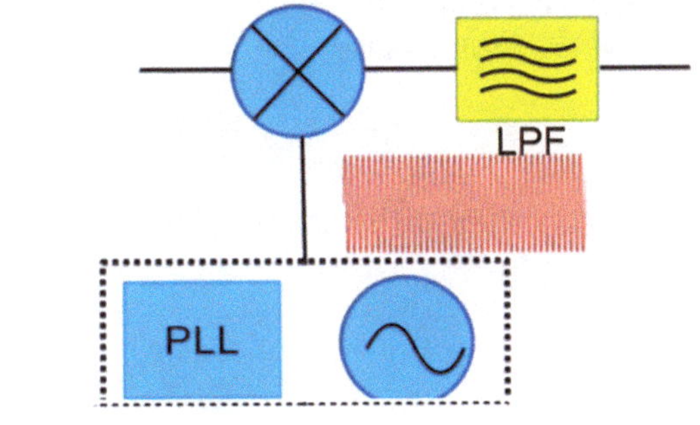

Fig. 7.11 The configuration of LC high pass filter

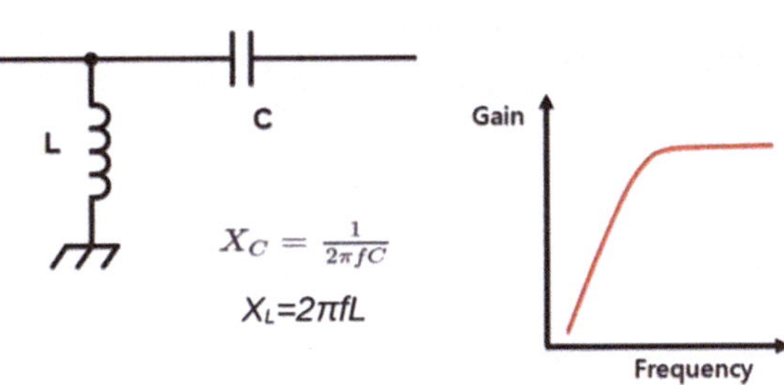

but they are also grounded by that capacitor so they are not passed to the output. Now, those low pass filters are widely used in the RF circuits and they are used to block the high frequency noise and they are also used in demodulation procedure because we know at the receiver or in demodulator, we have a lowpass filter (LPF) that only allows the demodulated message signal to pass through as illustrated in Fig. 7.10.

7.7 The LC High Pass Filter

Now, the LC high pass filter is used to block the low frequency noise in the audible frequency ranges. The audible frequency ranges are from 0 to 20 kHz, and if we have noise in the low frequencies, we can use the HPF to remove that noise. We have an LC high pass filter in Fig. 7.11, and the inductor is connected in parallel and the capacitor is connected in series, so we adjust the value of inductance in such a way that for the higher frequencies we want to pass that inductor has a very high reactance, or in other words, for those frequencies that inductor acts as an open circuit but for the low frequencies we want to block and for those frequencies that inductor reacts as a short circuit, so those low frequencies are grounded by that inductor, similarly for the capacitor we choose the capacitor in such a way that for low frequencies that we want to block that capacitor has a very high reactance or in other words for those low frequencies that capacitor acts as an open circuit and does not allow those frequencies to pass through but for the higher frequencies that capacitor has a low reactance and allows those higher frequencies to pass to the output so in that way we implement the LC high pass filter.

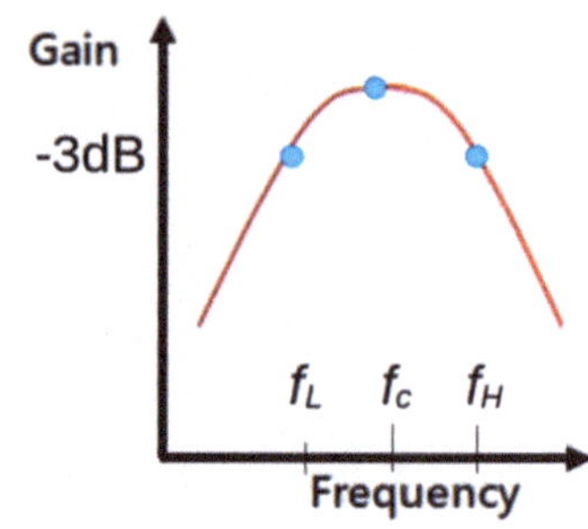

Fig. 7.12 The frequency response of a bandpass filter

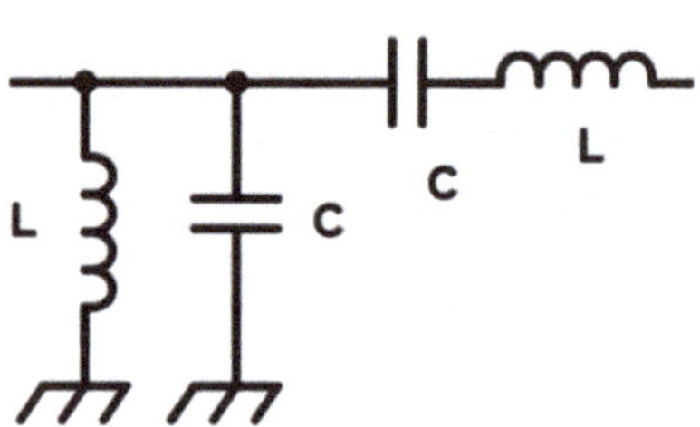

Fig. 7.13 The LC bandpass filter (BPF) circuit

7.8 The Q-Factor of a Filter

Now, we want to define the Q factor of a filter, the frequency response of a bandpass filter (BPF) is shown in Fig. 7.12.

The center frequency is f_c while the f_L is the lower 3 dB cutoff frequency and f_H is the higher 3 dB cutoff frequency and Q factor is defined by the following equation.

$$Q = \frac{f_c}{f_H - f_L} \tag{7.4}$$

If the bandwidth ($f_H - f_L$) of that BPF is small in that case we will have a high value of Q factor but if the bandwidth of that filter is high then we will have a low Q factor value.

7.9 The LC Bandpass Filter

The LC bandpass filter passes only signals at a specific frequency range and cuts or blocks signals at other frequencies, and it can be made using two LC circuits, so it consists on an LC tank circuit that is connected in parallel and an LC series circuit that is connected in series as depicted in Fig. 7.13.

Those two LC circuits have the resonant frequency that is equal to the center frequency f_c of the bandpass filter so for that parallel LC tank circuit would offer a very high impedance at the center frequency and the frequencies around it so those frequencies would path through and would not be grounded by the LC tank circuit, however for other frequencies

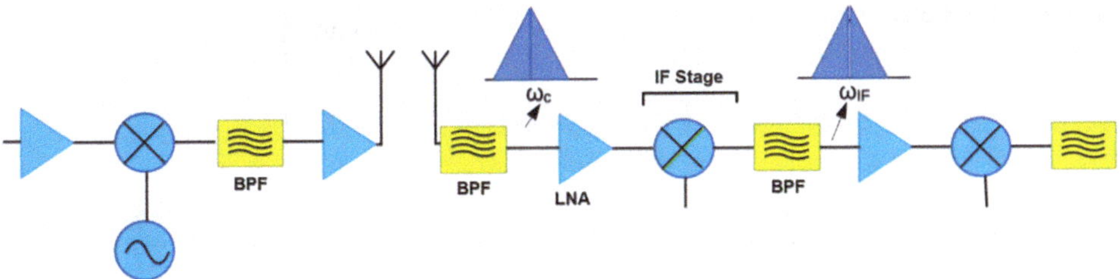

Fig. 7.14 The LC bandpass filter (BPF) usage

Fig. 7.15 The LC circuit for band stop filter

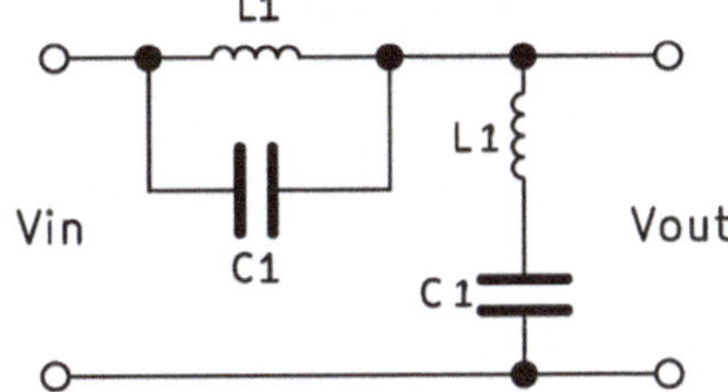

that parallel LC tank circuit would have a low impedance and for those frequencies that parallel LC tank would act as a short circuit and those frequencies would be grounded. Now, for the LC series circuit, for the center frequency f_c and around it that LC series circuit would offer a very low impedance and those desired frequencies they would pass through it while for the other frequencies that LC series circuit would offer a very high impedance and those frequencies would be blocked. Now, how can we increase the bandwidth (f_H-f_L) of the LC bandpass filter with the frequency response as demonstrated in Fig. 7.12? In order to increase the bandwidth of that BPF we need to increase the Q factor of that filter and that Q factor can be increased or adjusted using a resistance in series to inductor in the parallel LC tank circuit. Now, this LC bandpass filter is used in the transmitter in order to let the modulated signal to pass through and it is used at the receiver in order to let the received modulated signal pass through while rejecting the noise and other frequencies and it is also used after the intermediate frequency (IF) stage in order to filter the message signal that is at the intermediate frequency so, the LC BPF is used in modulation procedure and before LNA and also after IF stage as illustrated in Fig. 7.14.

7.10 The LC Band Stop Filter

The LC band stop filter is used to stop a particular range of frequencies and the LC circuit for band stop filter is shown in Fig. 7.15.

As you can see in Fig. 7.15, it uses two LC circuits, one is the LC tank circuit that is connected in series and the other is the LC series circuit that is connected in parallel and

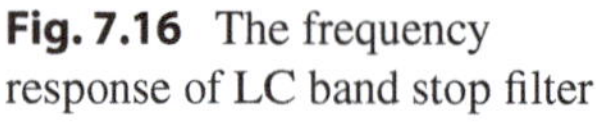

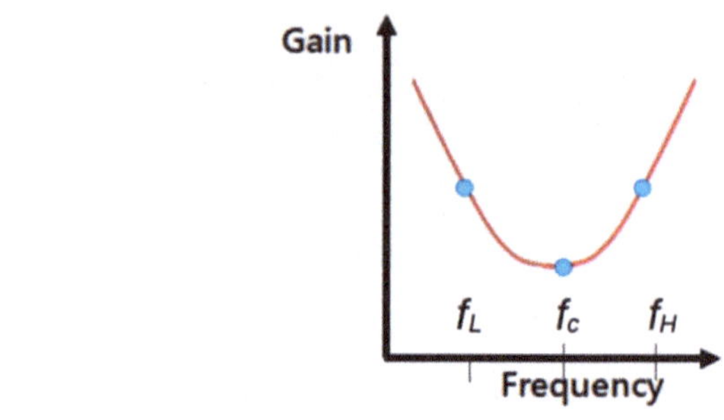

Fig. 7.16 The frequency response of LC band stop filter

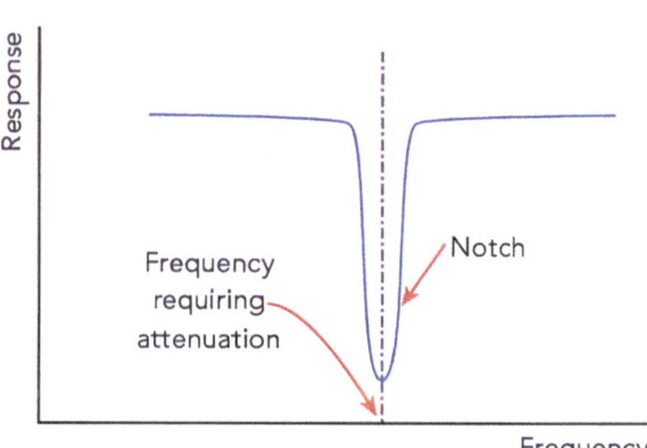

Fig. 7.17 The notch filter frequency response

those LC circuits have their resonant frequencies set at the center frequency fc of band stop filter. Now, for that center frequency and the frequencies around it that LC tank circuit would offer a very high impedance and it would stop those frequencies, while at the same time that LC series circuit would offer a very low impedance to those frequencies and they would be grounded by that LC series circuit so by the combined action of that LC tank circuit and the LC series circuit those frequencies would not reach the output, while for the other frequencies that LC tank circuit would offer a very low impedance and they would pass through and they would be offered a very high impedance by the LC series circuit so they would not be grounded and those frequencies would reach the output as depicted in Fig. 7.16.

Now, the notch filter is a special type of band stop filter that has a very narrow bandwidth and it is used to filter out the particular frequency as illustrated in Fig. 7.17, for example the notch filter can be used to filter out the AC power supply noise that exists at 60 Hz.

7.11 The Active Lowpass Filter

The active filters use op-amps to attenuate the selected frequencies and to amplify the desired frequencies. An example of the active lowpass filter is shown in Fig. 7.18.

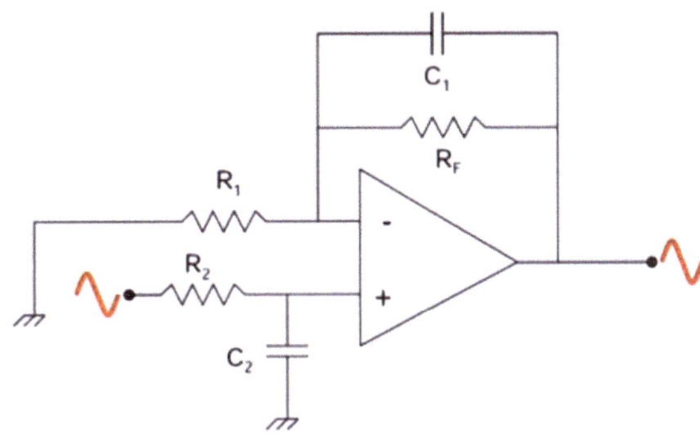

Fig. 7.18 An example of the active lowpass filter

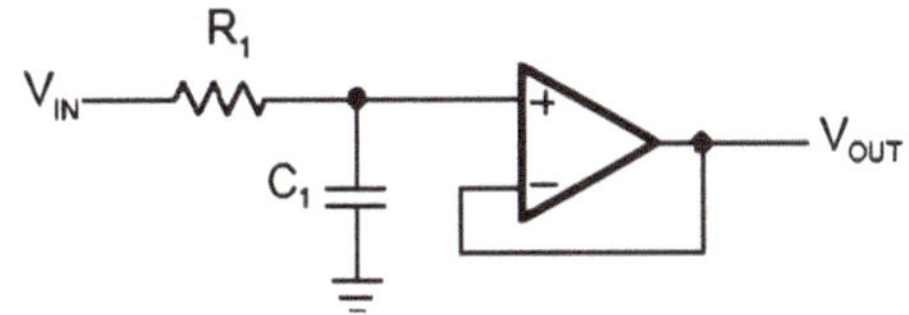

Fig. 7.19 The first order lowpass filter

As you can see in Fig. 7.18, the input is being applied to the non-inverting input of the op-amp and that input is passing through a filter that consists of resistance R_2 and capacitor C_2, and for higher frequencies the capacitor C_2 acts as a short circuit and that means those higher frequencies are grounded by that capacitor and they do not reach the input of the op-amp and at the same time for higher frequencies the capacitor C_1 acts as a short circuit so there is a higher negative feedback coming to the op-amp for the higher frequencies, so there is a very less gain for higher frequencies. So, by the combined actions of those two filters, the input filter does not allow the higher frequencies to pass through and at the same time there is a very less gain for the higher frequencies, so in this way the higher frequencies are blocked while for the lower frequencies they pass through the input filter and at the same time for the lower frequencies the capacitor C_1 has a very high impedance and that means there is a very less negative feedback for the lower frequencies, so the lower frequencies are amplified more, so in this way that circuit acts as an active lowpass filter.

7.12 The Order of a Filter

The order of a filter depends upon the number of the reactive elements in the filter. For example, in the lowpass filter in Fig. 7.19 we have one reactive element which is a capacitor so the order of the filter is one.

While in the filter as depicted in Fig. 7.20, we have two capacitors or in other words we have two reactive elements so the order of that filter is two.

In Fig. 7.21 we have cascaded two op-amps and in that filter, we have three reactive elements or three capacitors so the order of the filter is three.

Fig. 7.20 The second order lowpass filter

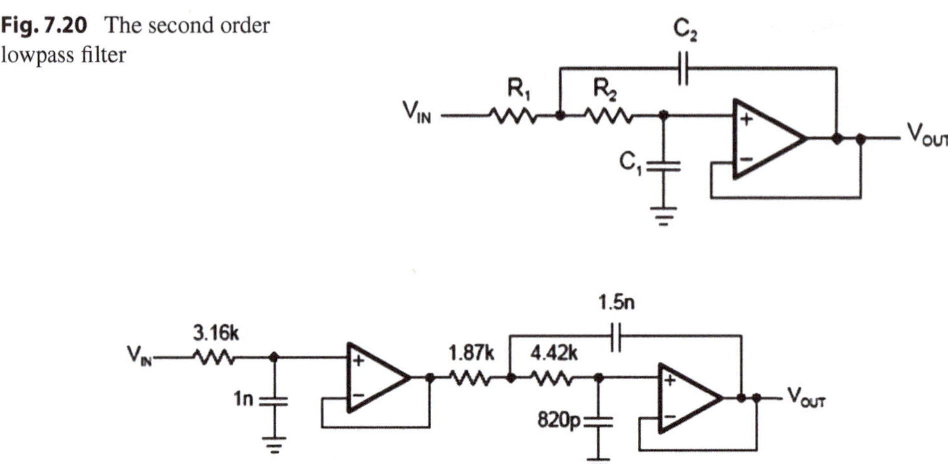

Fig. 7.21 The cascaded third order lowpass filter

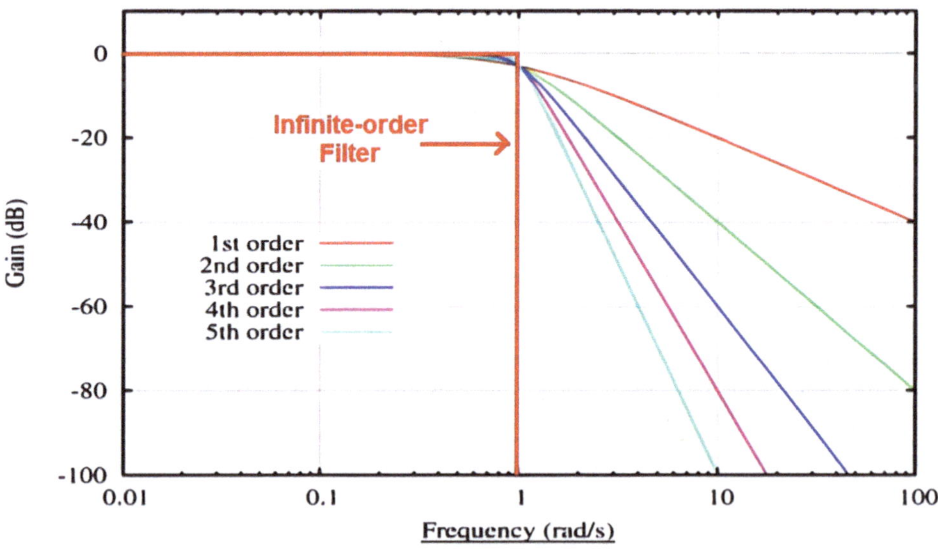

Fig. 7.22 The sharper frequency response for higher order of the filter

Now, what is the effect of increasing the order of the filter? When we increase the order of the filter, the filter keeps on getting sharper as illustrated in Fig. 7.22.

If we have an infinite order filter, then in that case that filter becomes an ideal lowpass filter as demonstrated in Fig. 7.22. However, we know that in order to construct an infinite order filter we need infinite reactive elements which is not possible in the real world.

7.13 Conclusion

This chapter has provided a thorough exploration of Radio Frequency (RF) filters, the critical circuits responsible for signal integrity and spectral management in communication systems. Beginning with their fundamental purpose—to pass desired frequencies and reject all others—we established the four primary filter types (LPF, HPF, BPF, BSF) and their vital applications in both transmitter and receiver chains.

A key portion of the chapter established the essential vocabulary for quantifying RF systems, detailing the use of decibel-milliwatts (dBm) for power levels and decibels (dB) for gain, attenuation, and the 3-dB bandwidth that defines a filter's practical passband. This analytical framework is indispensable for system design.

The operational principles of core passive LC filter topologies—lowpass, highpass, bandpass, and band-stop—were analyzed, explaining how inductors and capacitors combine to create frequency-selective behavior. We further defined the Quality factor (Q) as a measure of selectivity and filter order as the determinant of roll-off steepness, connecting circuit complexity to performance.

The chapter concluded by introducing active filters, which integrate operational amplifiers to provide gain and overcome the insertion loss limitations of purely passive networks, representing a more versatile design approach.

In summary, this chapter has equipped you with the principles to understand, specify, and analyze the filters that shape the frequency spectrum. Mastery of these concepts—from basic LC circuits to the metrics of dB and Q—is foundational for designing the amplifiers, mixers, and complete RF systems that will be discussed in the following chapters.

References

1. Tiwana M (2021) RF concepts, components and circuits for beginners. Udemy Inc., San Francisco, CA
2. Ludwig R, Bretchko P (2000) RF circuit design: theory and applications. Pearson, Upper Saddle River, NJ

The RF Power Amplifiers

8

Contents

8.1 Introduction to the Power Amplifiers

In this chapter we will discuss about the power amplifiers, the power amplifier is shown in Fig. 8.1 and that takes the modulated signal, amplifies it and sends it to the antenna so that it can be transmitted over the wireless channel.

Now, that power amplifier operates at a higher frequency and at a high power so its design is much more challenging when it is compared to the amplifier that is before the mixer and since that amplifier is operating at a lower frequency and at a lower power so compared that amplifier the design of that power amplifier is much more challenging. Now in general, the amplifiers are divided into classes A, B, C, D, E, F and so on and those amplifier classes can be used from very low frequencies to the microwave range and now

M. Pakdel, *Understanding RF Systems*, Synthesis Lectures on RF/Microwaves,
https://doi.org/10.1007/978-3-032-19227-1_8

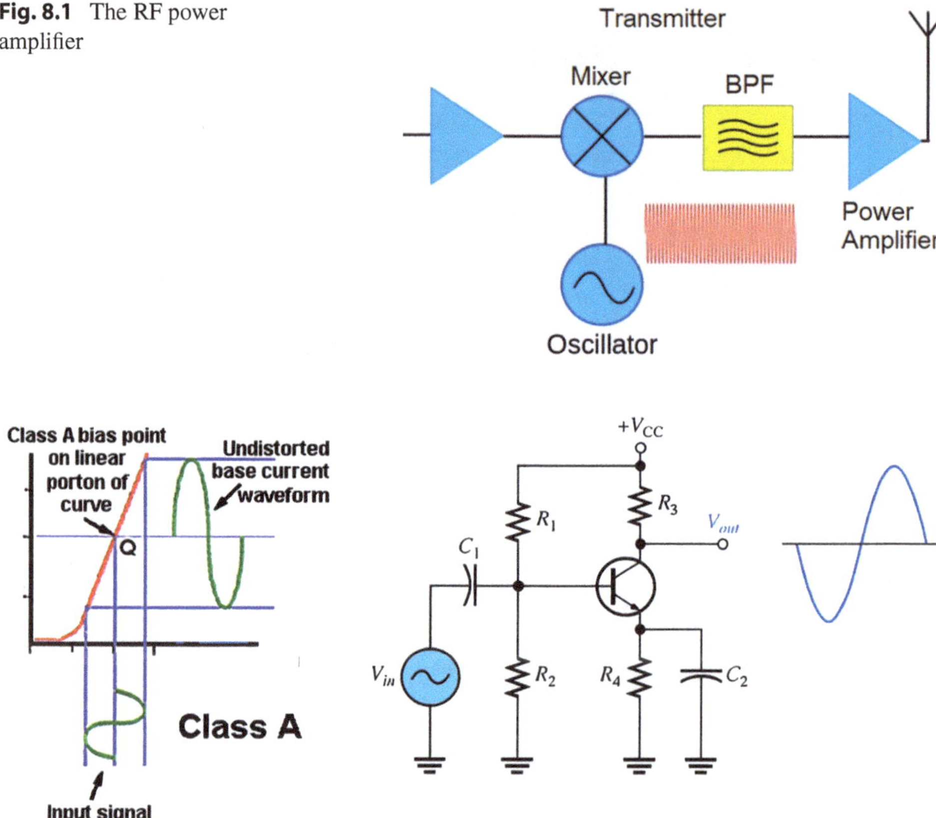

Fig. 8.1 The RF power amplifier

Fig. 8.2 The class A amplifier characteristics

we are going to discuss the important amplifier classes and what are the pros and cons of those amplifier classes [1].

8.2 The Class A Amplifiers

Now first we would discuss the class A amplifiers and the class A amplifier is depicted in Fig. 8.2. As you can see in Fig. 8.2, it is using a transistor in the common emitter (CE) configuration and we are applying the input signal that we need to amplify to the base of the transistor and at the output (collector) we are getting the amplified signal. The resistances are used for the biasing of the transistor, in other words we can say that those resistances set the operating point of the transistor for example in Fig. 8.2, the Q is the operating point of the transistor and the red line indicates the relationship between the input and output signals, and if there is no input signal ($V_{in} = 0$), in that case that transistor or amplifier would be operating at the operating point of Q, and when the input signal is applied

Fig. 8.3 The 8-QAM signal

there we can see that it is amplified by the transistor or amplifier and at the output we get the amplified signal. Also, the shape of the input signal waveform and the shape of the output signal waveform are the same, so there is no distortion. So, we can say that amplifier has a high linearity and that is possible because that transistor is conducting for the full 360° cycle of the input signal. However, that amplifier has a poor power efficiency (PE), because when we are not applying any input, even in that case that amplifier is operating at that point Q and it is drawing the current from V_{CC}.

We calculate the power efficiency (PE) by the following equation.

$$PE = \frac{P_{out}}{P_{dc}} \tag{8.1}$$

We get the output power (P_{out}) from the signal at the collector of transistor and the dc power (P_{dc}) is the power we are supplying to the circuit for the operation of that circuit and the downside of this amplifier is that it has a low power efficiency [2].

8.3 The Linear Amplification Necessity

So, why do we require the linearity in an amplifier? The linear amplification is required when we are using the amplitude modulation or a combination of both amplitude and phase modulation schemes which is QAM scheme, and an example of 8-QAM is illustrated in Fig. 8.3.

As we can see in Fig. 8.3, there is variation in the phase as well as the amplitude of the modulated signal so we need amplify it in such a way that there is a very little distortion in that signal because the information is also contained in the amplitude of the signal, so for such modulation schemes we require the linear amplification. However, on the other hand the modulation schemes like the frequency modulation (FM) schemes where the information is carried in the form of variation in the frequency or the phase modulation (PM) scheme where the information is carried in the form of variation in the phase but they have the constant amplitude, for those modulation schemes we can use the amplifiers that are not very linear as shown in Fig. 8.4.

8.4 The Class B (Push-Pull) Amplifiers

Now, we discuss about the class B amplifiers which are also called as the push-pull amplifiers and this amplifier consists of two transistors as depicted in Fig. 8.5.

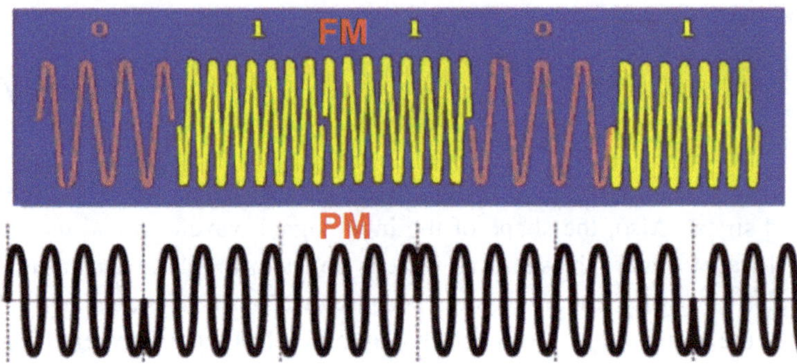

Fig. 8.4 The frequency and phase modulation (FM and PM) schemes

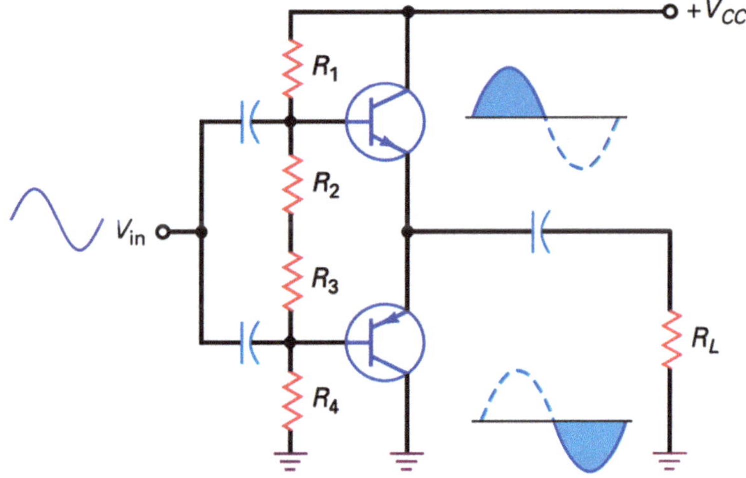

Fig. 8.5 The class B (push-pull) power amplifier

So, during the positive half of the input signal cycle the upper transistor is conducting and it is pushing the current to the resistance R_L down to the ground and during the positive half of the input signal cycle, the lower transistor is conducting and it is pulling the current through the resistance R_L from ground to upward. So, each transistor operates for 180° cycle. The due line that indicates the relationship between the input and the output is illustrated in Fig. 8.6.

As you can see in Fig. 8.6 for the voltage of 0.7 V, the upper transistor is not conducting, and also for −0.7 V the lower transistor is not conducting so as a result if we see the output at the point of zero crossing of the signal there is distortion because the transistors are not conducting at those points. So, these amplifiers suffer from crossover distortion, and we can see if there is no input signal, the amplifier would be operating at the origin point so that means that would be drawing a very little current from the power supply V_{CC}. So, as a result these amplifiers have a high efficiency up to 78.5%, but they have the

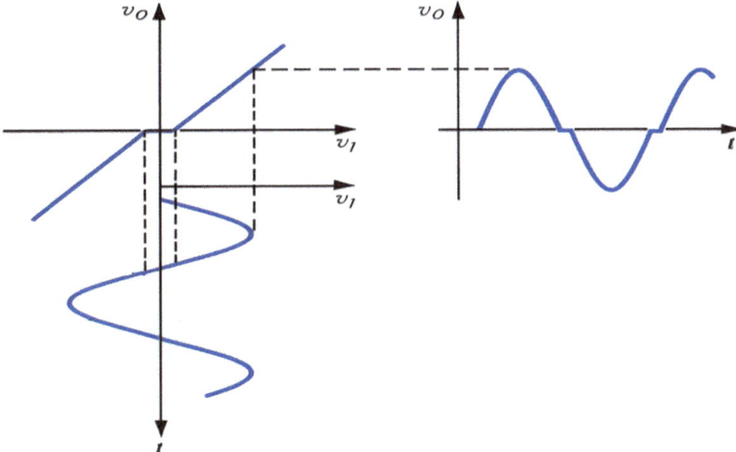

Fig. 8.6 The due line indicating the input and output relationship

problem of crossover distortion so these amplifiers are used in the low power RF devices where the quality of voice is not a big concern for example in the case of pocket radios.

8.5 The Class AB Amplifiers

Now, we discuss about the class AB amplifiers, the class AB amplifiers are a compromise between the class A that have good linearity and class B amplifiers that have a good efficiency and if we see the circuit of class AB amplifier then we can see that it is similar to push-pull or the class B amplifier but in this case we have put an additional diode that provides an additional 0.7 V input the upper transistor and also we have put another diode that provides additional −0.7 V input to the lower transistor so in this way those transistors are not completely cut-off during zero crossing of the input signal and in this way the crossover distortion is avoided and so we can say that the conduction angle of each transistor is likely above 180° and it is between 180° and 360°. So, these amplifiers have a good efficiency between 50% and 78.5% and at the same time they have a good linearity and little crossover distortion as shown in Fig. 8.7.

8.6 The Class C Amplifiers

Now, we discuss about the class C amplifiers. The circuit of class C amplifier is depicted in Fig. 8.8.

As you can see in Fig. 8.8, at the input of the transistor we have a resistance R_1 which is called as the pull-down resistor because it is used to pull down the operating point of the transistor as illustrated in Fig. 8.9.

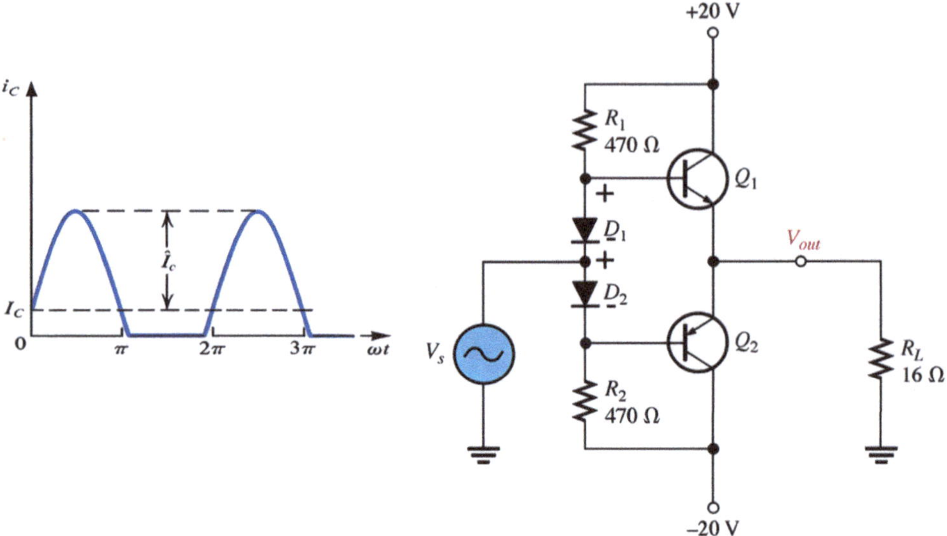

Fig. 8.7 The class AB amplifier

Fig. 8.8 The circuit of class C amplifier

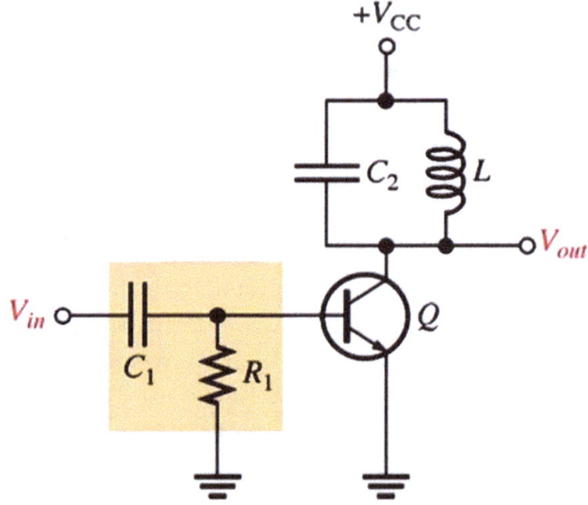

The input and output waveforms are demonstrated in Fig. 8.9, and we can see from the output waveform that the transistor conducts for a very small cycle which is less than 180° and as a result the output current would be as shown in Fig. 8.10.

As you can see in Fig. 8.10, there is a heavy distortion at the output current, now that output current of the transistor is used to trigger the LC tank circuit which resonates at the frequency that is set by the value of L and C_2 and as a result we get a high frequency oscillation at the output so this amplifier is used as an RF oscillator and that amplifier is not used in the audio applications to amplify the audio signal because of the distortion

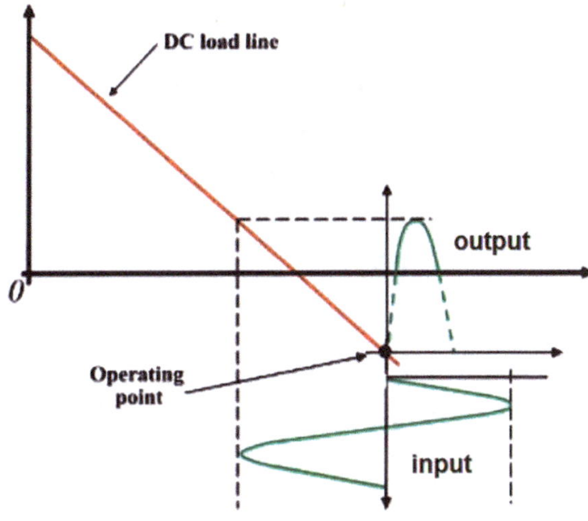

Fig. 8.9 The operating point of the transistor

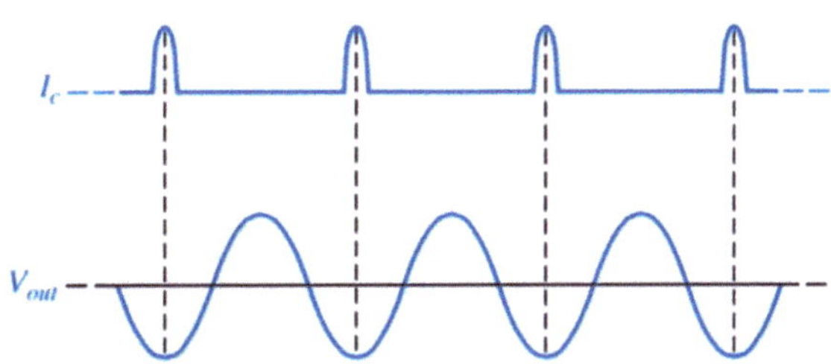

Fig. 8.10 The output current and voltage waveforms

problem. Now in those amplifiers since the transistor operates for a small conduction cycle so those amplifiers have a very high efficiency about 90% but those amplifiers suffer from heavy distortion problem at the output.

8.7 The Additional Classes of Amplifiers (D, E, F, and So on)

There are other classes of amplifiers as well like D, E, F and so on and these amplifiers use the transistor as a switch and the conduction cycle of these transistors or these amplifiers is very small so they have a very high efficiency which is near 100% and those amplifiers also suffer from very high non-linearity so these amplifiers are only used in very specialized applications so that is why we are not going to discuss these amplifiers further as depicted in Fig. 8.11.

8.8 The 1 dB Compression Point as the Linearity Measure

The curve in Fig. 8.12 shows the relationship between the output and input of an amplifier.

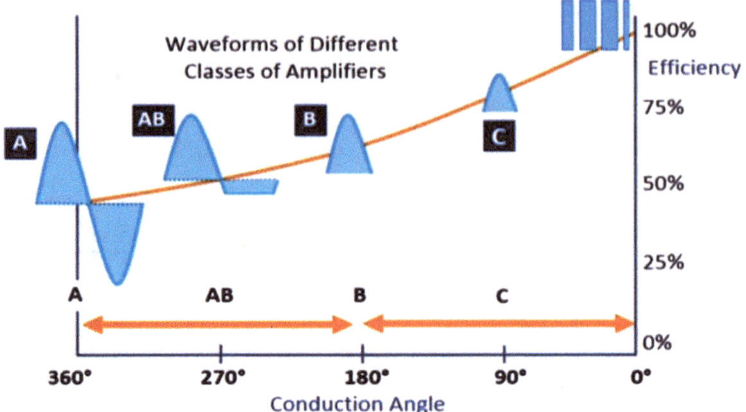

Fig. 8.11 The amplifier classes characteristics

Fig. 8.12 The relationship between the output and input of an amplifier

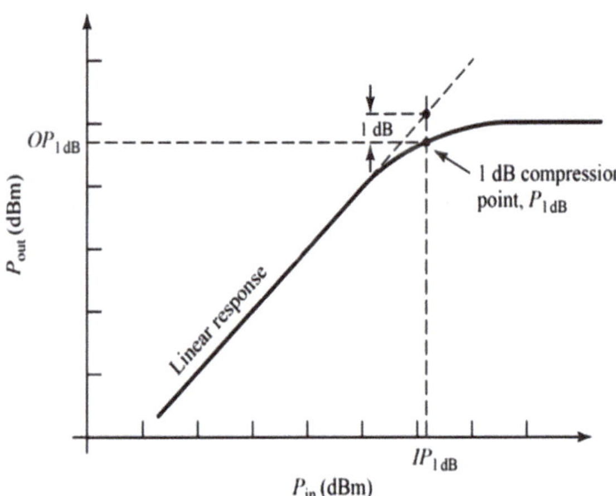

As you can see in Fig. 8.12, the input power of the amplifier is P_{in}(dBm) and as we increase the input power of the amplifier, we can see the output power (P_{out}(dBm)) of the amplifier is also increasing linearly in that region up to a point and after that point the nonlinearity begins and the gain of the amplifier start decreasing and then a point comes after which that gain is flatten out and the amplifier is completely saturates and now increasing the input power does not increase the output power so ideally we would have wanted that the input-output curve of an amplifier is like that linear dashed line as demonstrated in Fig. 8.12 an it has the total linear region of operation but in reality the curve is nonlinear. Now, the point where the difference between two curves is 1 dB we called it as the 1 dB compression point (P_{1dB}) and it is the measure of the linearity of the amplifier because more that point Is towards the right, the more the longer is the linear response region of the amplifier and the better the amplifier is, so 1 dB compression point is the measure of linearity.

8.9 The Amplifier Backoff Operation

Now, when the amplifier is operating at 1 dB compression point, it has the maximum power efficiency because after that point the gain starts decreasing and at that 1 dB compression point we are getting the maximum gain and that means for that input power, we are getting the maximum output power as illustrated in Fig. 8.13.

So, we have optimum power efficiency, however that point is the non-linear point of operation of the amplifier and we want to avoid non-linearities. So, what is the solution for that problem? The solution of that problem is to reduce the input power so that the amplifier operates in the linear region but close to the non-linear region, now that reducing of the input power of the amplifier is called the backoff operation of the amplifier and that difference between the 1 dB operation point and the actual point of the operation of that amplifier is measured in dBs and it is called as the backoff of the amplifier as demonstrated in Fig. 8.13.

8.10 The Intermodulation Due to Nonlinearity

Now, what happens when we operate the amplifier in the non-linear region of its operation? In that case the harmonics would be produced. For example, if we are operating an amplifier in the non-linear region of operation and we are inputting two frequencies f_1 and f_2 to that amplifier then the harmonics of those frequencies would be formed for example in the case of f_1 the harmonics $2f_1$, $3f_1$, $4f_1$ and so on would be formed and in the case of frequency f_2, the harmonics $2f_2$, $3f_2$, $4f_2$ and so on would be formed and those harmonics would add and subtract from one another to form the intermodulation products. For

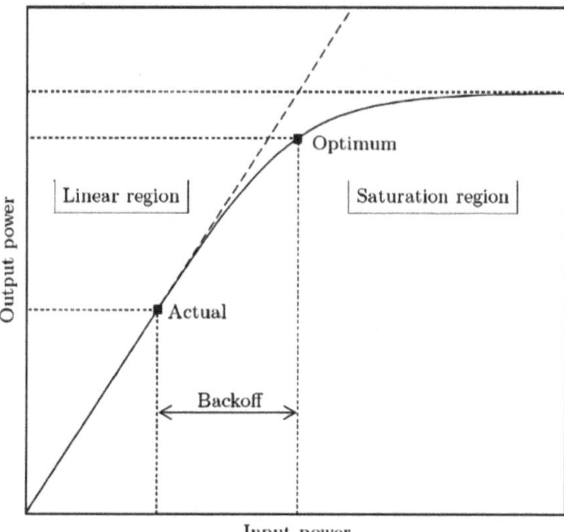

Fig. 8.13 The amplifier Backoff operation

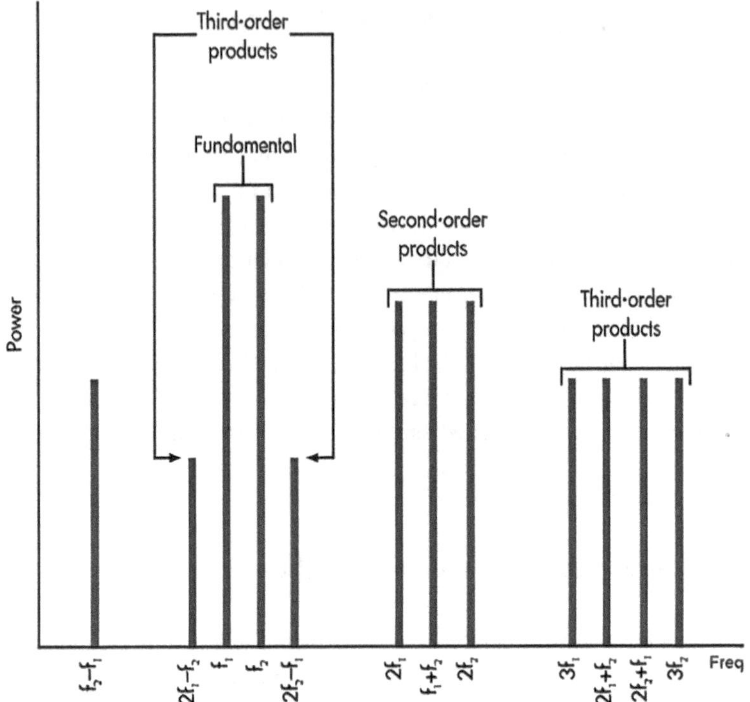

Fig. 8.14 The intermodulation (IM) components

example, when we are talking about the second order harmonics, we have $2f_1$, $2f_2$, but in addition to that we have $f_1 + f_2$ and f_1-f_2 as the second order harmonics and in the case of third order harmonics, we have $3f_1$, $3f_2$, and we have sum and difference of them $2f_1 \pm f_2$ and $2f_2 \pm f_1$ as the third order harmonics as shown in Fig. 8.14.

So, in Fig. 8.14 we have different intermodulation products or different harmonics so we have second order products, third order products, fourth order products and so, now the most problematic are a part of third order products which are $2f_1-f_2$ and $2f_2-f_1$ because the rest of intermodulation products can easily be filtered out using a filter but it is difficult to filter out those two intermodulation products which are third order and they are very near to our input signal f_1 and f_2 as demonstrated in Fig. 8.14.

8.11 The Third Order Intercept as the Linearity Measure

Now, we have seen that the third order harmonics or the third order intermodulation products are the most problematic and the input-output curve of the third order harmonics in Fig. 8.15 shows that for a given input power of a signal what is the power of the third order harmonics at the output and also there is a cure that shows for a given power input what is

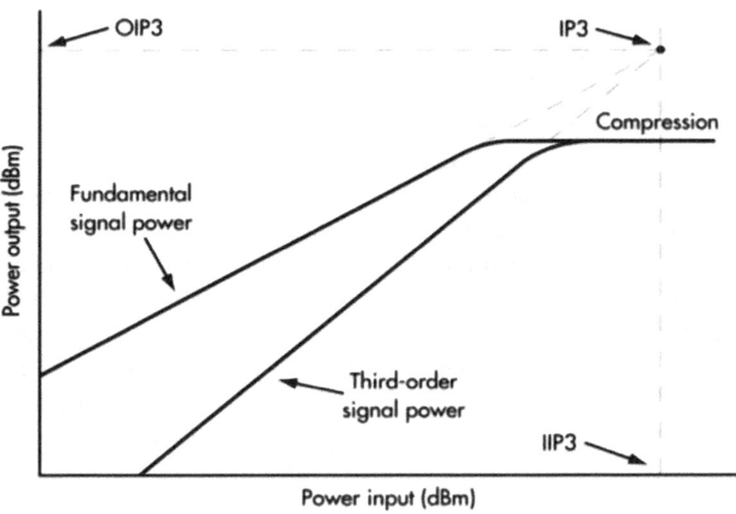

Fig. 8.15 The third order intercept as the linearity measure

the output power of our desired signal and we always want that the power of third order products is lower than the power of our actual signal and that means we want that the lower cure is lower than the upper curve.

Now we can see in Fig. 8.15 that those two curves have the linear tendency up to their 1 dB points and after that those two curves saturate out, however if we extend the linear lines of those two curves, then those two linear lines would meet at point IP3 which is called as the third order intercept. Now if IP3 is located towards the right then it is better or the amplifier is better so that means that for a significant long region of operation of the amplifier the power of third order products is lower than the power of our actual desired signal so that is the measure of the linearity of the amplifier we want that third order intercept of amplifier is located towards the right as much as it is possible.

8.12 Conclusion

This chapter has provided a comprehensive exploration of Radio Frequency (RF) Power Amplifiers, the critical final-stage components that determine the transmitted signal's power, efficiency, and spectral integrity. We began by outlining the distinct challenges of PA design, which must operate at high frequencies and power levels where performance trade-offs become acute.

A systematic analysis of core amplifier classes revealed the fundamental compromises between linearity and efficiency. We examined Class A for its high linearity, Class B and AB for their improved efficiency with managed distortion, and Class C for its very high efficiency in non-linear applications, linking each class's suitability to specific modulation schemes like QAM or FM.

The chapter then established the essential analytical framework for evaluating and specifying PA performance in real-world systems. Key metrics such as the 1-dB Compression Point (P1dB) and the Third-Order Intercept Point (IP3) were introduced as crucial measures of linearity and distortion. We explained the practical technique of amplifier backoff to manage the linearity-efficiency trade-off and analyzed the generation of problematic intermodulation distortion (IMD), particularly third-order products.

In summary, this chapter has equipped you with the principles to understand the design choices, operational limits, and performance specifications of RF power amplifiers. Mastery of these concepts—from class topology to linearity metrics—is essential for designing efficient, compliant RF transmitters, forming a direct link to the broader system design and signal integrity challenges addressed in subsequent chapters.

References

1. Pozar DM (2011) Microwave engineering, 4th edn. John Wiley & Sons, Hoboken, NJ
2. Tiwana M (2021) RF concepts, components and circuits for beginners. Udemy Inc., San Francisco, CA

The Antennas

9

Contents

9.1 Introduction to Antenna and Its Polarization

In this chapter we will discuss about the antennas, an antenna is a device that converts electrical signal into electromagnetic waves and also it does the reverse procedure of converting the electromagnetic wave back into the electrical signal for example when we talk about the transmitting antenna, the transmitting antenna converts the electrical signal into electromagnetic waves and those waves are then radiated over the wireless channel as shown in Fig. 9.1.

M. Pakdel, *Understanding RF Systems*, Synthesis Lectures on RF/Microwaves,
https://doi.org/10.1007/978-3-032-19227-1_9

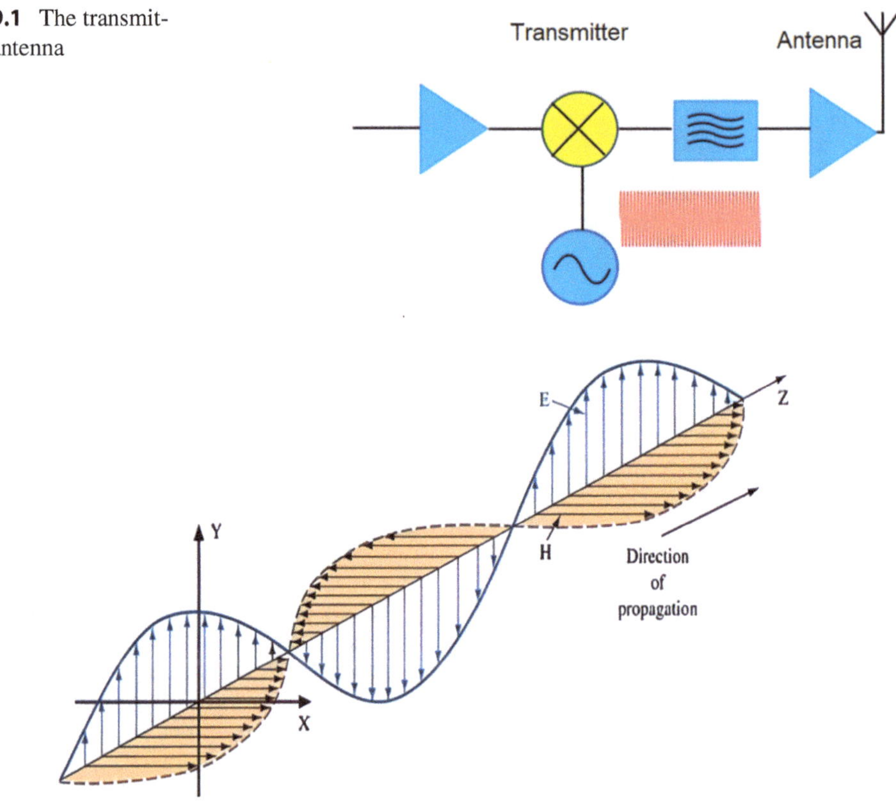

Fig. 9.1 The transmitting antenna

Fig. 9.2 The electromagnetic wave and its direction of propagation

Similarly, the receiving antenna receives the electromagnetic waves from the received beam into electrical signal. The electromagnetic wave and direction of propagation of electromagnetic wave is depicted in Fig. 9.2.

As you can see in Fig. 9.2, the electromagnetic wave consists of two fields, one is the electrical filed (Y axis) and to the 90° angle of electrical field we have the magnetic field (X axis), now the polarization of an antenna is the direction of its electrical field. For example, if we have placed the antenna in the vertical direction, the electrical field that would be produced it would also have the vertical polarization and to receive such vertically polarized electromagnetic wave we would need an antenna that has also the vertical polarization at the receiver as illustrated in Fig. 9.3.

Similarly, if at the transmitter we have used an antenna that has horizontal polarization and it generates horizontal polarized electromagnetic wave in order to receive it at the receiver we also need a horizontally polarized antenna as shown in Fig. 9.4.

However, there is a special type of polarization which is called as the circular polarization that can be generated using the special antenna which is called as the helical antenna and that antenna produces a circular polarized electromagnetic wave and the beauty of this

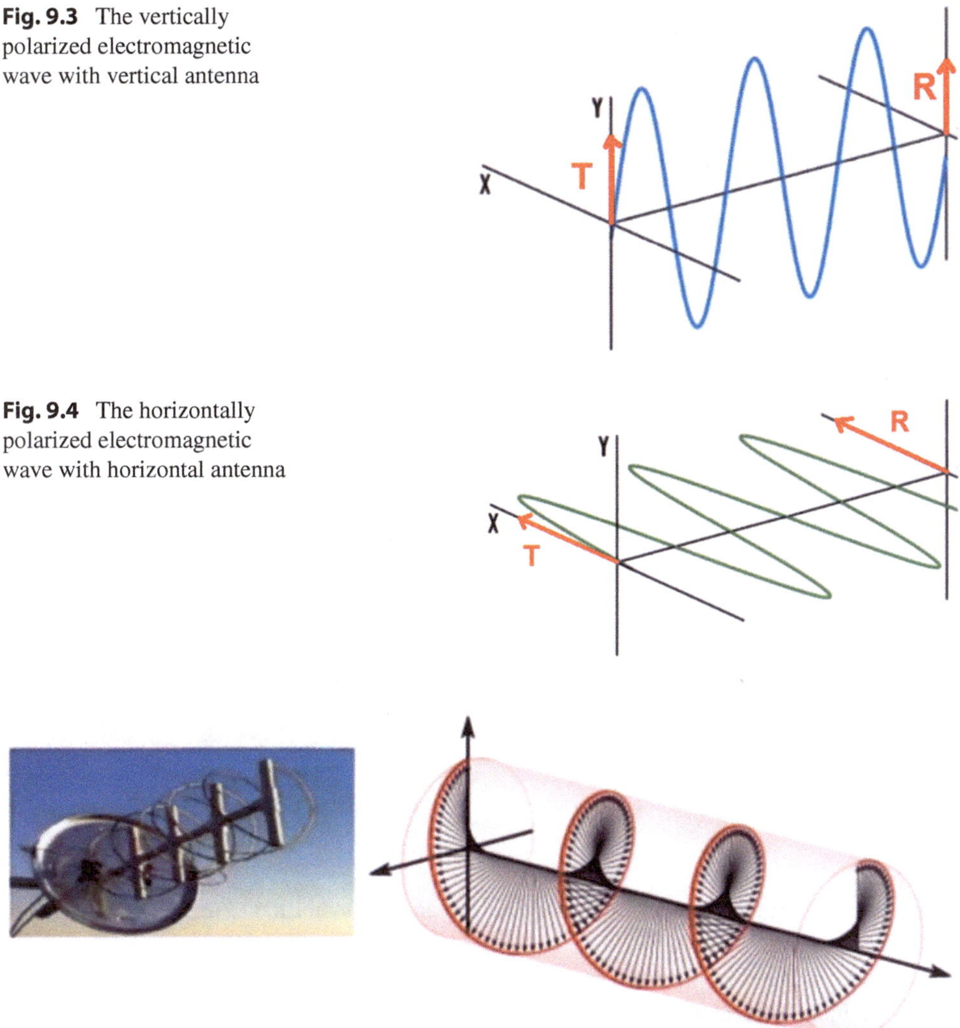

Fig. 9.3 The vertically polarized electromagnetic wave with vertical antenna

Fig. 9.4 The horizontally polarized electromagnetic wave with horizontal antenna

Fig. 9.5 The circular polarization for both vertical and horizontal antennas

electromagnetic wave is that at the receiver whether we place our antenna vertically or horizontally in both the cases the transmitted signal can be received at the receiver as depicted in Fig. 9.5 [1].

9.2 The Important Antenna Parameters

So, what are the important parameters of an antenna? The first one is the center frequency or center wavelength of the signal that can be transmitted to this antenna and the second important parameter is the bandwidth around the center frequency or the range of

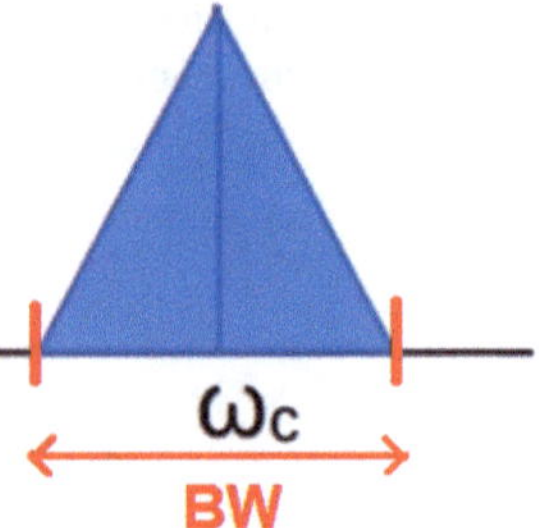

Fig. 9.6 The signal center frequency and bandwidth

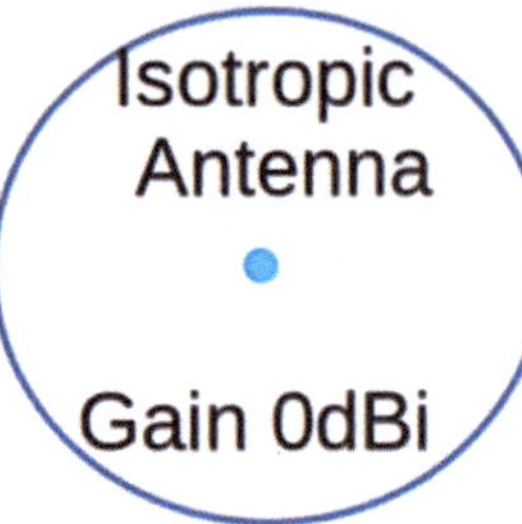

Fig. 9.7 The isotropic antenna

Fig. 9.8 The directional antenna

frequencies around that center frequency which can be transmitted to this antenna as illustrated in Fig. 9.6.

The other important parameters are gain and the pattern of the electromagnetic beam that is produced by that antenna and we are going to discuss these two parameters in more detail in this chapter.

9.3 The Antenna Gain

Now, we discuss about the antenna gain, the antenna is a passive device and that means the power that is radiated by an antenna can never be greater the power that is input to an antenna. In Fig. 9.7 we have the example of isotropic antenna and the isotropic antenna is a hypothetical antenna that does not exist in practice and it consists of a point antenna that radiates equally in all directions.

When we talk about the directional antennas, the directional antennas have the ability to focus their electromagnetic energy in one particular direction as shown in Fig. 9.8 and we define the gain of that directional antenna as the power transmitted by that directional antenna in direction of its maximum focus divided by the power transmitted by the isotropic antenna in that direction.

When we convert the ratio of those two powers into dBs, we get the gain of the antennas in dBi and here the "i" stands for isotropic because the isotropic antenna is being used as the reference and that gain is called as the isotropic antenna gain. Now in the case of isotropic antenna when we are comparing the isotropic antenna with itself in that case it is quite natural that its gain would be OdBi as demonstrated in Fig. 9.7 [2].

9.4 The Dipole Antenna

Now, the simplest antenna that we can make it is the dipole antenna and in Fig. 9.9 we have a transmitter that generating an electrical signal that we want to transmit using antenna and that electrical signal is being carried to the antenna using the pairs of cables which is called as the transmission line.

The simplest antenna that we can make is by bending that transmission line at its ends so in Fig. 9.10 we have bend that transmission line at its ends in that form and the total

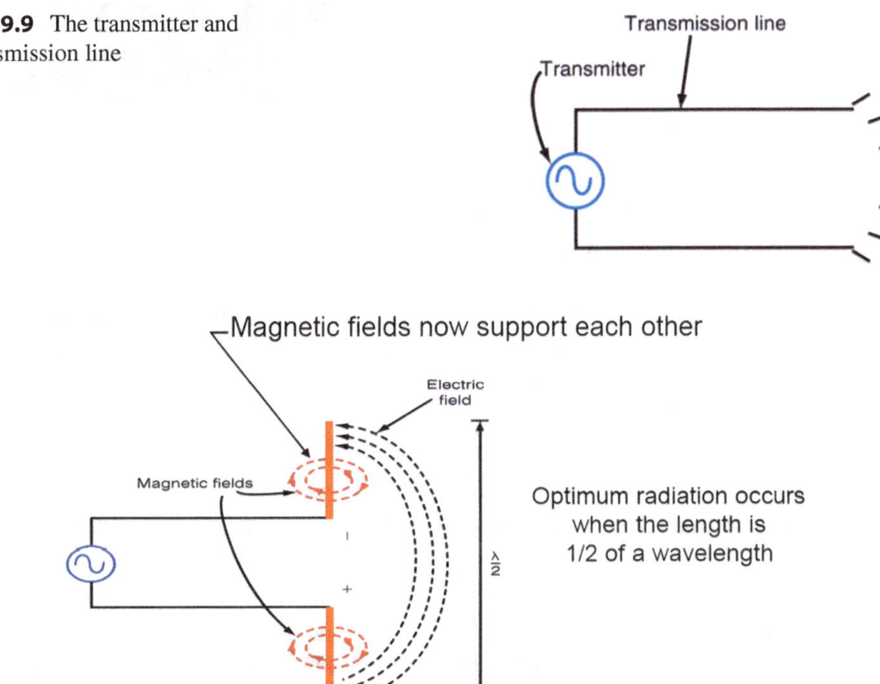

Fig. 9.9 The transmitter and transmission line

Fig. 9.10 Bending the transmission line at its ends (half wave dipole antenna)

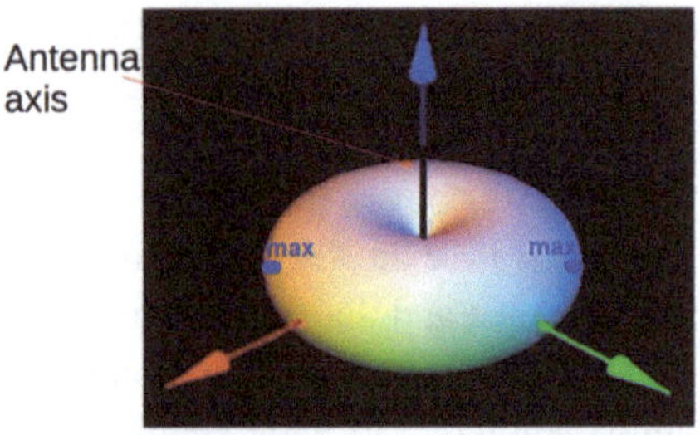

Fig. 9.11 The radiation pattern of dipole antenna

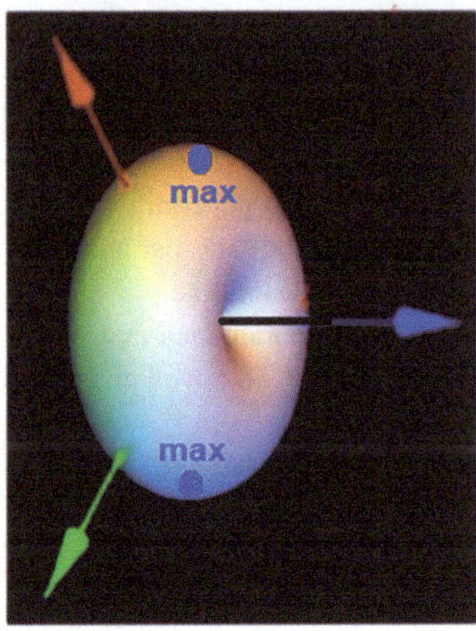

Fig. 9.12 The rotation of the diagram in Fig. 9.11

length of the bend transmission line is equal to λ/2, where the λ is the center wavelength of the signal that we want to transmit using the antenna.

Now, in the configuration demonstrated in Fig. 9.10 the electrical field that is formed by those two ends of the antenna it would reinforce one another and as a result it excesses in antenna and it transmits the electrical signal as the electromagnetic signal over the wireless channel and since the length of the antenna is equal to λ/2, that is why it is also called as the half wave dipole and it is one of the most widely used antennas because of its simplicity.

9.5 The Dipole Antenna Radiation Pattern and Beamwidth

Now, if we place the dipole antenna in the vertical axis, then the radiation pattern would be produced as shown in Fig. 9.11.

That means we would get the maximum radiation power of the electromagnetic signal in the directions as demonstrated in Fig. 9.11 and if we rotate the diagram in Fig. 9.11 and take the cross section of that diagram as depicted in Fig. 9.12, then we would get the diagram as illustrated in Fig. 9.13.

The point where the maximum radiation of the electromagnetic signal would take place is demonstrated in Fig. 9.13 and as we move towards left or towards the right of that point the power of the electromagnetic radiation would decrease and at another point the power of the electromagnetic radiation would reduce to the half value of the maximum power that is why we also call it as the 3 dB down point from the maximum radiation, so we have also 3 dB down point and if we measure the angle between those two 3 dB down points, we would get the beamwidth of that dipole antenna so that is how the beamwidth of a dipole antenna is defined.

9.6 The Monopole (Marconi) Antenna

The monopole antenna is also called as the Marconi antenna and this antenna is similar to the dipole antenna except that one end of the transmission line is grounded and this ground acts as a reflecting mirror for that upper half of the antenna that has the length of λ/4. So,

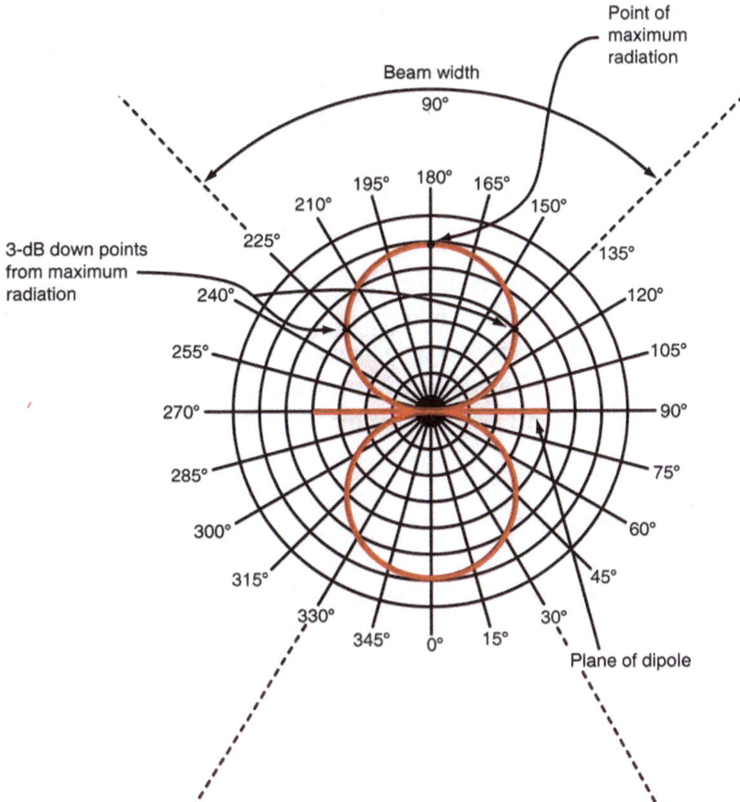

Fig. 9.13 The cross section of dipole antenna radiation pattern

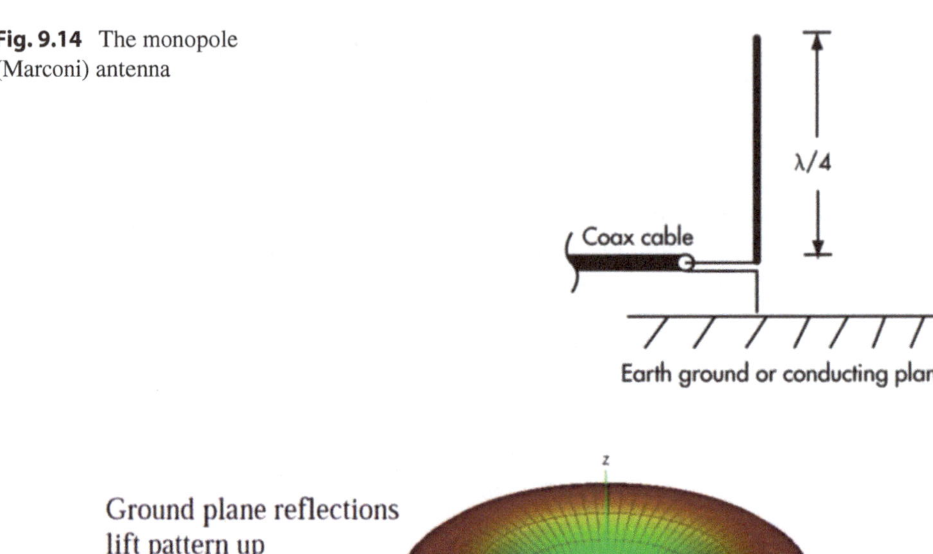

Fig. 9.14 The monopole (Marconi) antenna

Fig. 9.15 The radiation pattern of the monopole antenna

monopole has half the length of the dipole antenna because that ground plane is acting as the mirror for the upper plane as shown in Fig. 9.14.

The radiation pattern of the monopole antenna is similar to the dipole antenna except due to the reflection from the ground plane it is lifted as depicted in Fig. 9.15 and the gain of the monopole antenna is 1 dB less than the dipole antenna.

9.7 The Loop Antenna

The examples for the loop antenna are illustrated in Fig. 9.16.

As you can see in Fig. 9.16, we have two examples for the loop antenna and the left hand side loop antenna is used for the RFID chip and we can see that this antenna is in the form of a coil that is forming a loop and that coil is carrying the electrical signal that needs to be radiated as the electromagnetic signal and that antenna may be circular, rectangular, triangular, square or hexagonal in shape and the length of that loop is about the wavelength of the signal that we want to transmit. So now that is not a good idea to have many turns in the loop in order to reduce the size of that antenna because when we do those turns there

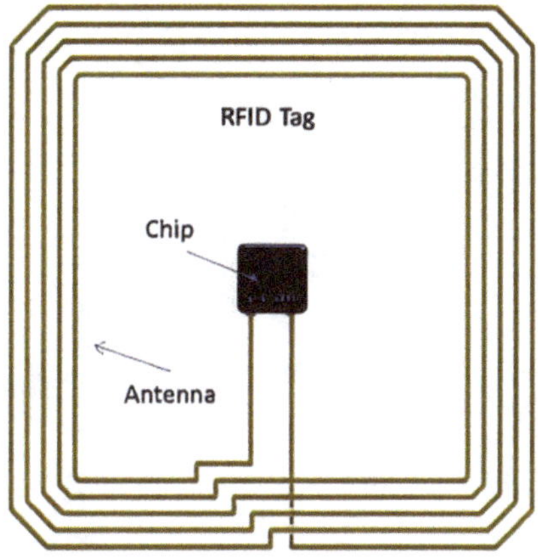

Fig. 9.16 The examples of the loop antenna

can be capacitance between those turns and this can result in the low efficiency of the antenna.

9.8 The Travelling Wave Types Antenna

These antennas are called as the traveling wave type antennas because the RF currents in these antennas generates radio waves that travel through the antenna in one direction only and they are different from resonant antennas that we discussed earlier like monopole or dipole in which the antenna acts as a resonator and the current travels or bounces back and forth from the ends of antenna to produce the standing waves and those standing waves then generates the electromagnetic radiations, now the advantage of non-resonant antennas or the traveling wave type of antennas is that they have a higher bandwidth as compared to the resonant antennas as shown in Fig. 9.17.

The first type of traveling wave type antenna that we are going to discuss it is the Yagi antenna, and it is depicted in Fig. 9.18.

As you can see in Fig. 9.18, in one end we have a reflector and then you have the driven element, and it is called as the driven element because it is there that you connect the electrical signal that you want to transmit and then you have the directors, you have multiple directors and if you want to increase the gain of the Yagi antenna then you add more directors so the gain of the Yagi antenna depends upon the number of directors and these directors are nothing but they are straight rods and that is why the construction of the Yagi antenna is very simple and it is very robust. The radiation pattern of the Yagi antenna is illustrated in Fig. 9.19.

As you can see in Fig. 9.19, one end is the directors end and the other end is the reflector end and one disadvantage of these antennas is that for high gains those antennas become very long as you have to add more directors and the maximum gain that can be

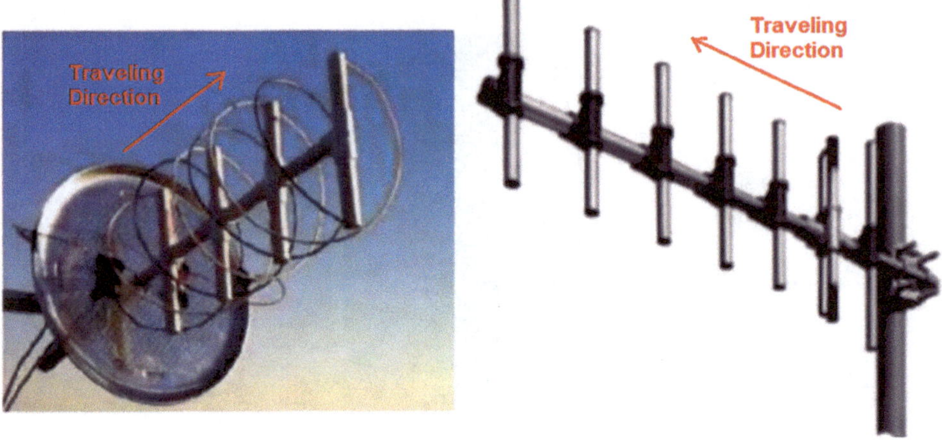

Fig. 9.17 The traveling wave type of antennas

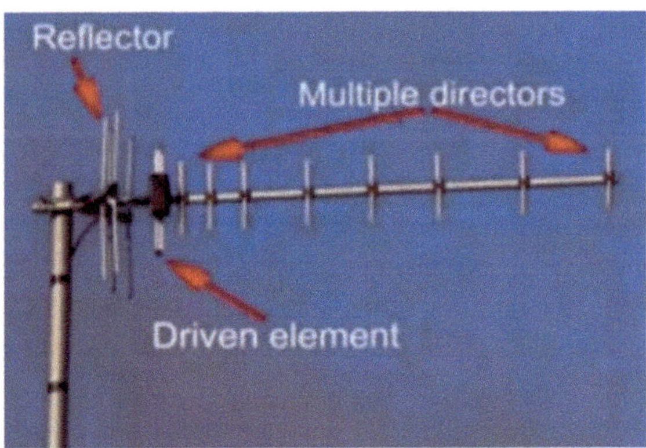

Fig. 9.18 The Yagi antenna

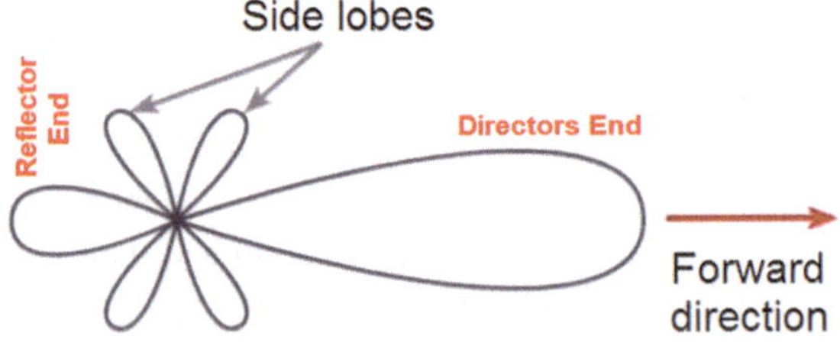

Fig. 9.19 The radiation pattern of the Yagi antenna

provided by these antennas it is 20 dB. The second type of traveling wave type of antennas is the helical antenna and, in this antenna, the conducting wire is bound in the form of a helical and that wire is connected to the ground plate, and that ground plate is then connected to the feeder as shown in Fig. 9.20.

The helical antenna is the simplest antenna that be used in order to produce circularly polarized waves, and that antenna is frequently used in extraterrestrial communication in

Fig. 9.20 The helical antenna

Fig. 9.21 The log-periodic antenna

order to communicate with the satellites and the space probes outside earth because we do not know what is the polarization of the antennas of those satellites or space probes.

9.9 The Log Periodic Antenna

The log-periodic antenna is a directional antenna that is designed to operate over a wide range of frequencies that means this antenna has a very high bandwidth and this antenna consists of half wave dipole driven elements that have the gradually increasing length as depicted in Fig. 9.21.

As we add more half wave dipole driven elements to that antenna the bandwidth of the antenna increases further and the distance between those elements is the logarithmic function of the frequency.

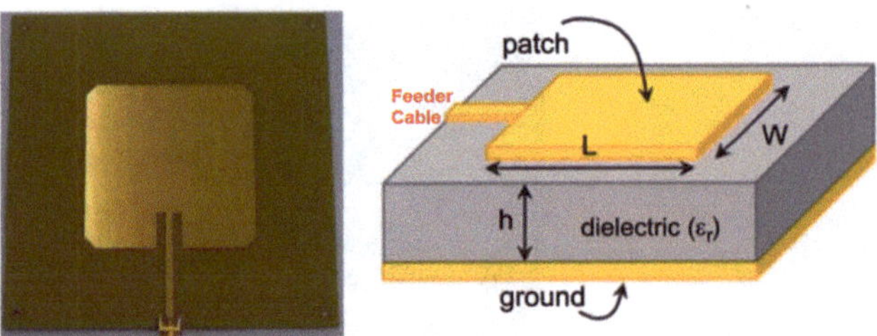

Fig. 9.22 The microstrip or patch antenna

Fig. 9.23 The shape of the horn antenna

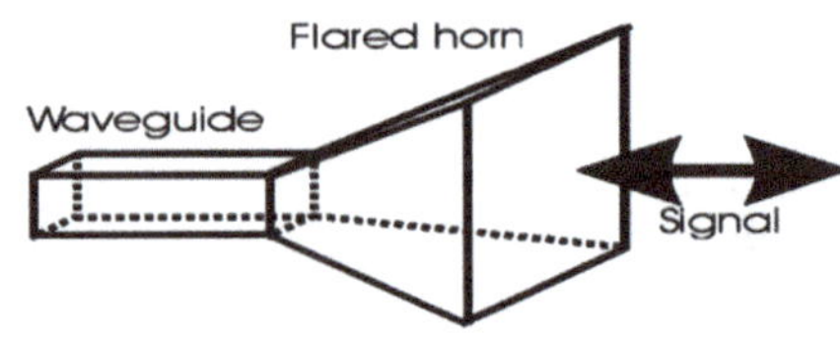

9.10 The Microstrip or Patch Antenna

Now, we discuss about the microstrip or patch antenna, this antenna consists of a very thin metallic strip which is placed on the ground plane with a dielectric material placed in between them as illustrated in Fig. 9.22.

As you can see in Fig. 9.22, the patch may be in square, or circular, or a rectangular shape because that eases our mathematical analysis and the fabrication of that microstrip antenna, and that microstrip and the feeder cable is etched on the dielectric using the photo-etching procedure and the length of the metal patch is $\lambda/2$. The disadvantages of that type of microstrip antenna are that it has a narrow bandwidth and it has an inefficient radiation.

9.11 The Horn Antenna

Now, we discuss about the horn antennas, the horn antennas are used for the frequencies of 4GHz and above and the shape of the horn antenna is shown in Fig. 9.23.

As you can see in Fig. 9.23, the horn antenna is fed the electromagnetic signal by the waveguide and that waveguide is then connected to the feeder cable. The horn antennas have a wide beam which is especially useful for the GEO satellites. The GEO satellites are 36,000 km away from the earth and they have the coverage area of about 1/3 of earth as depicted in Fig. 9.24.

So, that wide beam is especially useful for the GEO satellites and these horn antennas provide with the gain of about 20 dBi and have a beamwidth of 10° and higher and if a

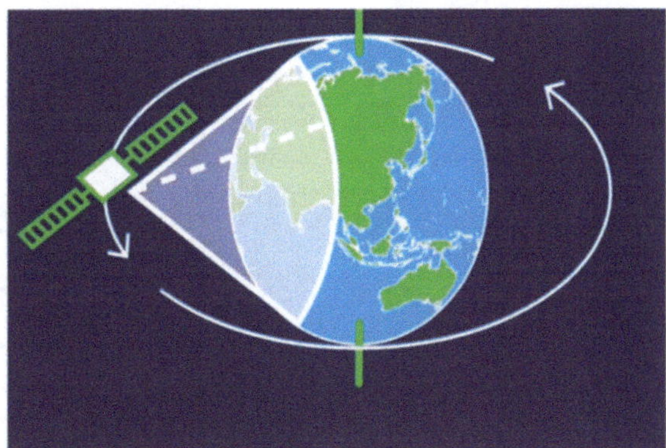

Fig. 9.24 The coverage area of GEO satellites

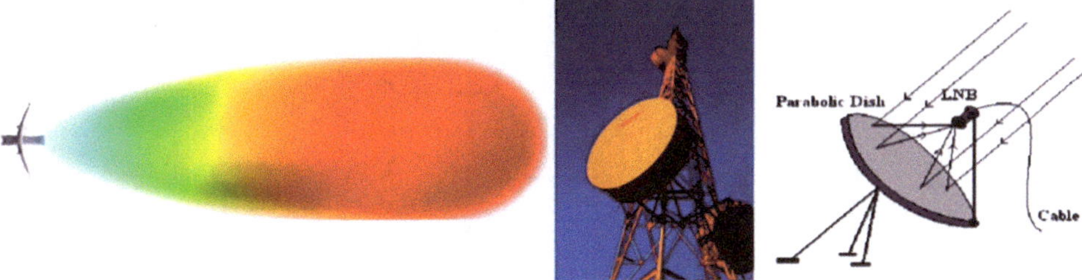

Fig. 9.25 The circular parabolic reflector antennas

narrower beam or a higher gain is needed then instead of a horn antenna a reflector or an array antenna is used.

9.12 The Circular Parabolic Reflector Antenna

Now, we discuss about the circular parabolic reflector antennas, so these antennas are most common type of antennas that are used for the point-to-point transmission links because they are easy to construct, they have a very good gain and a narrow beamwidth so they are good for many applications so as you can see in Fig. 9.25, the parabolic antenna has a very good beamwidth and a very good gain. Also, the structure of the parabolic antenna is demonstrated in Fig. 9.25 and if that is the transmitting parabolic antenna then we can see that the feeder cable is connected to the feed point and that feed point converts the electrical signal into the electromagnetic signal which is then reflected by the parabolic dish in a particular direction and in the case of receiving parabolic antenna the received electromagnetic signal is focused on the feed point from where it is converted back to the electrical signal and given to the cable.

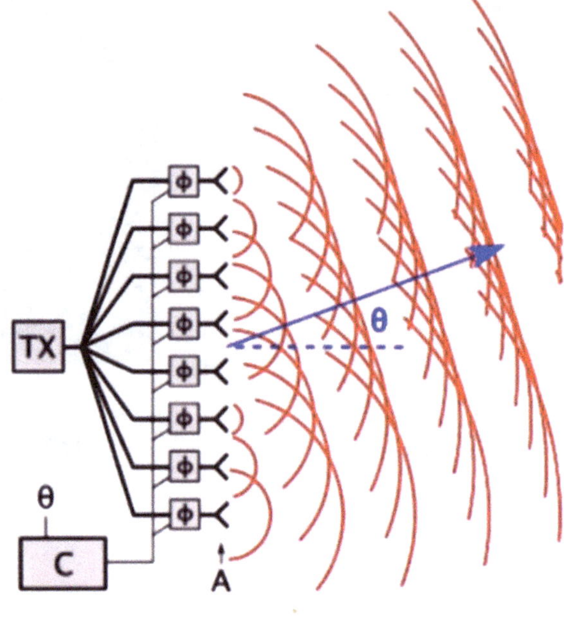

Fig. 9.26 The structure of array antenna

9.13 The Array Antennas

The array antenna consists of many small dipoles, helices or horns as its elements and those elements are connected to the RF input signal as illustrated in Fig. 9.26.

However, we adjust the amplitude and phase of each individual element in such a way that the electromagnetic waves that are produced they constructively interfere in one particular direction so the beam is formed in that direction and by adjusting the amplitude and the phase of the signal that is given to each individual element we can change the direction of the beam and we can also make that beam more sharper by adding more antennas in the array elements so no physical movement of the antenna system is necessary in order to steer the beam in the different directions.

9.14 Conclusion

This chapter has provided a comprehensive exploration of antennas, the fundamental components that bridge the electronic circuits of an RF system to the free-space wireless channel. We began by establishing the core function of an antenna as a transducer between guided electrical signals and radiated electromagnetic waves, introducing the critical concept of polarization and its role in ensuring efficient transmitter-receiver alignment.

The chapter established a framework for antenna analysis by defining key performance parameters: center frequency, bandwidth, gain (in dBi), and the radiation pattern. A systematic survey of common antenna architectures followed, revealing a progression from simple resonant structures to sophisticated wideband and electronically controlled designs. We analyzed fundamental resonant antennas like the half-wave dipole and monopole, defined their beamwidth, and explored compact variants like the loop antenna. The principles of traveling-wave antennas were then introduced, leading to the analysis of the directional Yagi-Uda, the circularly-polarized helical antenna for satellite links, and the broadband log-periodic antenna.

The chapter progressed to antennas critical for modern integrated and high-frequency applications, covering the microstrip patch antenna for compact devices and the horn antenna for microwave systems and satellite coverage. We concluded with high-performance directive systems: the parabolic reflector antenna for high-gain point-to-point links and the advanced phased array antenna, which enables electronic beam steering and shaping without mechanical movement.

In summary, this chapter has equipped you with the principles to understand, compare, and select antennas based on application requirements such as frequency, directivity, bandwidth, and physical constraints. The journey from the simple dipole to the electronically agile array illustrates the evolution of antenna technology, providing the essential link between the RF circuits discussed in previous chapters and the complete wireless communication system.

References

1. Tiwana M (2021) RF concepts, components and circuits for beginners. Udemy Inc., San Francisco, CA
2. Pozar DM (2011) Microwave engineering, 4th edn. Wiley, Hoboken, NJ

The Impedance Matching

10

Contents

10.1 Introduction to Impedance Matching

In this chapter we will discuss about the important concept of impedance matching, return loss and voltage standing wave ratio which is also called as the VSWR. Now what is the impedance matching? Suppose we have an RF device that is acting as a source and we have another RF device that is acting as a load and the output impedance of the source is 50 Ω and the input impedance of the load is also 50 Ω as shown in Fig. 10.1, in that case we can say that the impedances of the source and the load are matched to one another and in such condition the maximum power transfer between the source and the load would take place.

Now in the RF world the 50 Ω is the standard impedance that is used as the input and the output impedance of the RF devices but you may come across other systems as well where 75 Ω is used as the impedance standard for example in the case of cable TV system (CATV) [1].

M. Pakdel, *Understanding RF Systems*, Synthesis Lectures on RF/Microwaves,
https://doi.org/10.1007/978-3-032-19227-1_10

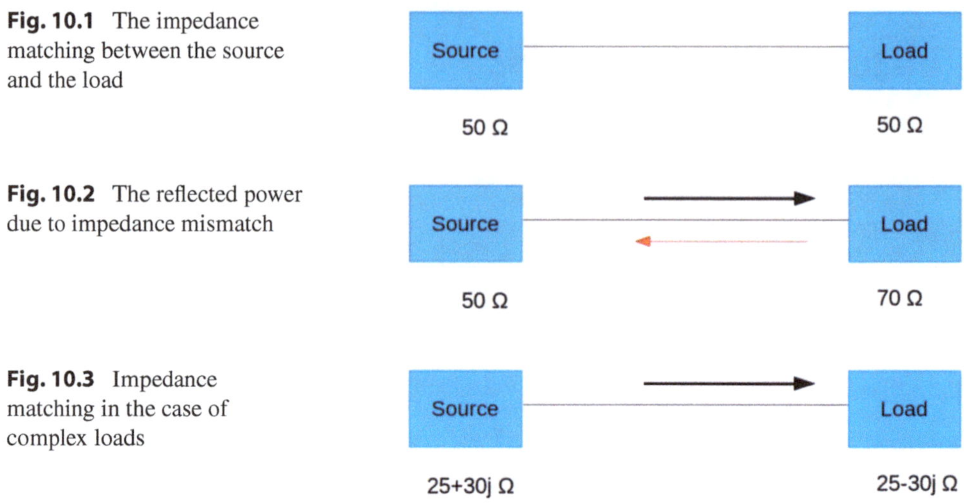

Fig. 10.1 The impedance matching between the source and the load

Fig. 10.2 The reflected power due to impedance mismatch

Fig. 10.3 Impedance matching in the case of complex loads

10.2 The Impedance Mismatch

Now, what will happen if the source and the load impedances are not matched? For example, if the output impedance of source is 50 Ω and the input impedance of load is 70 Ω in that case the impedances of source and load are not matched so what happens is that the power is being transferred from the source to the load, a part of that power is reflected back and that is called as the reflected power or the reverse power and that reflected power is undesirable in the RF circuits as depicted in Fig. 10.2.

10.3 The Impedance Matching in Complex Loads

Now, suppose we have a load that has a complex impedance, for example it has the impedance of 25-30jΩ where the -30j is the imaginary part of the impedance that is related to the reactive part of the load, in such case in order to match the source impedance with the load impedance the output impedance of the source must be the complex conjugate of the load impedance and in this case the complex conjugate of 25−30jΩ is 25 + 30jΩ, and we can see that the negative sign of imaginary part in the load impedance has been changed with the positive sign in the source impedance and in such condition we can say that the source and the load impedances are matched to one another as illustrated in Fig. 10.3 [2].

10.4 The Impact of Frequency on Reflected Power

Now, suppose we have a load that is purely resistive in nature, in that case when we change the frequency of the signal that is being transferred from the source to the load the reflected power does not change rather it remains constant and the example of purely resistive load

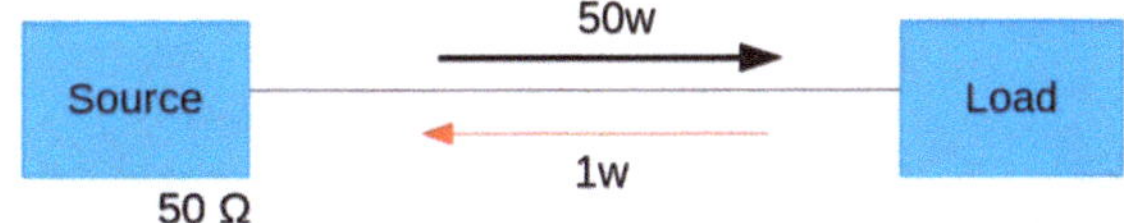

Fig. 10.4 The impact of frequency on reflected power

is the dummy load that is not changed with the frequency. Now if we have a complex load that has a reactive portion as well in that case increasing the frequency or changing the frequency also changes the amount of reflected power and increasing the frequency means that more power would be reflected from the load as shown in Fig. 10.4.

10.5 The Quantifying of Reflected Power

Now, in order to measure the reflected power, first we need to quantify the reflected power, and that reflected power is quantified as the parameters of return loss and the voltage standing wave ratio (VSWR). So, what is the return loss? The return loss is measured in dBs and it is the difference between the forward power that is going from source to the load which is measured in dBm and the reflected power in dBm that is being reflected by the load to the source and it is defined by the following equation.

$$Return\ loss(dB) = Forward\ power(dBm) - Reflected\ power(dBm) \tag{10.1}$$

Now, we can see from Eq. (10.1) that if the reflected power is smaller, then the return loss is larger so more return loss is always desirable. Then you have another parameter which is called as the VSWR. As you can see in Fig. 10.5, the green waveform represents the signal that is going from source to the load and the red waveform represents the signal that is being reflected by the load to the source, now if we add those two signals then we get another signal which is of blue color and that signal is in the form of a standing wave and we can see that blue standing wave has a maximum value which is being indicated by the V_{max} and it has a minimum value that is indicated by the V_{min} and if we divide the V_{max} by V_{min} then the ratio we get it is called as the voltage standing wave ratio (VSWR) as depicted in Fig. 10.5.

We can easily see from Fig. 10.5 that if the reflected power is higher, then the VSWR is higher and if the reflected power is lower the VSWR is lower.

10.6 Calculation of VSWR

Now, the VSWR can easily be measured using a device which is called as the network analyzer and the VSWR can also be calculated mathematically but for that first we have to calculate the reflection coefficient which is represented by the symbol of Γ and it is calculated by the following equation.

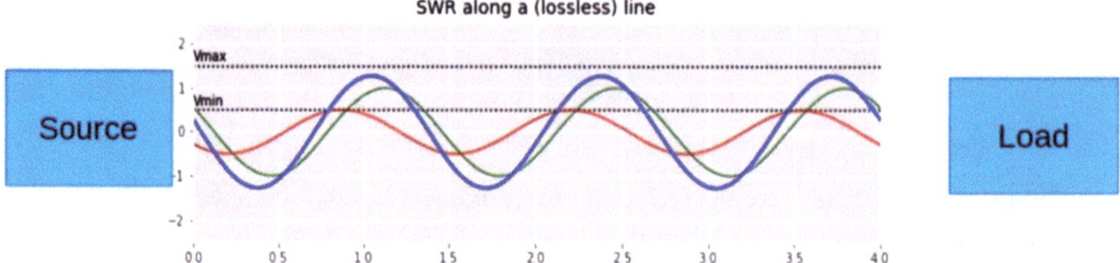

Fig. 10.5 The VSWR along a lossless line

Table 10.1 The VSWR Vs returned power

VSWR	Returned Power (approximate)
1:1	0%
2:1	10%
3:1	25%
6:1	50%
10:1	65%
14:1	75%

$$= \frac{Z_L - Z_o}{Z_L + Z_o} \tag{10.2}$$

Where, Z_L is the load impedance and Z_o is the source impedance, and once we get the Γ, we can easily use the following formula in order to calculate the VSWR.

$$VSWR = \frac{1+| \ |}{1-| \ |} \tag{10.3}$$

10.7 The VSWR and Returned Power

The Table 10.1 and the graph in Fig. 10.6 show the relationship between the VSWR and the returned power or the reflected power.

As you can see in Table 10.1, when the VSWR is 1 then the source and the load are perfectly matched and the returned power is 0%, but when VSWR is 2, in that case 10% of the power is reflected back to the source and when VSWR is increased to 3 the reflected power increases to 25% and when VSWR increases to 6, the reflected power increases to 50% and 50% returned power is very bad and this can really damage the source and then for the VSWR value of 10, 65% of power is returned and for the VSWR value of 14, 75% of power is returned and if we see the graph in Fig. 10.6, we can view that initially in the region of VSWR between 2 and 7, increasing the VSWR results in exponential increase in the reflected power, so it is really important to match the source and the load with one

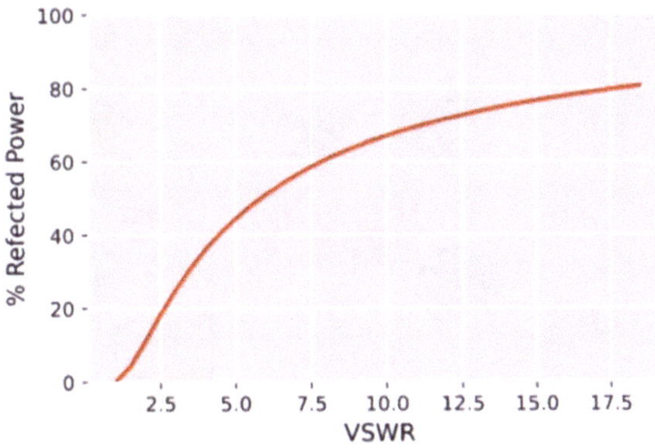

Fig. 10.6 The graph of VSWR Vs returned power

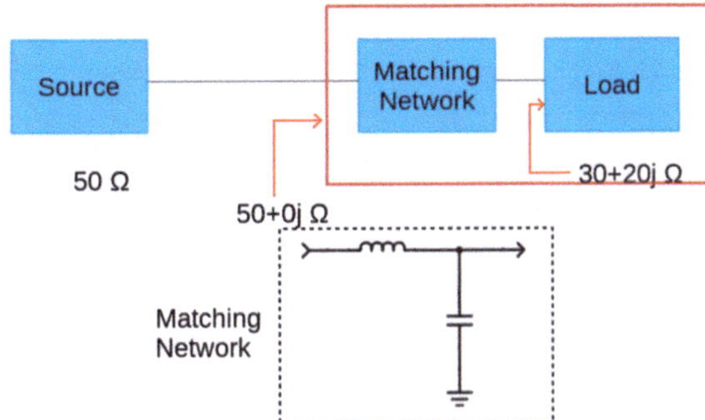

Fig. 10.7 The example of matching network

another so that this VSWR is minimized so that the reflected power is minimized. Now it is important to mention for the two special cases for example in the case of a short circuit and in the case of an open circuit and in both the cases 100% of power is reflected back to the source.

10.8 The Impedance Matching Networks

Now, in order to reduce the VSWR we use the matching network, for example in Fig. 10.7 we have a load that has a complex impedance of 30 + 20jΩ and if we see from the input of the load point towards the load the impedance od the load is 30 + 20jΩ and it does not match with the standard impedance of the 50 Ω that is there in the RF circuit so we use that matching network and that network for example may consist of an inductor and a capacitor and it changes the impedance of the load in such a way that when we see the impedance from the input of the matching network point, the overall impedance is 50 Ω which is matched to the impedance of the source which is also 50 Ω.

Now, the antenna is an example of the reactive load and different antenna types have different impedances and we know that in the RF circuit the standard impedance is 50 Ω so the standard impedance of the RF circuit and the devices in the RF circuit is 50 Ω but

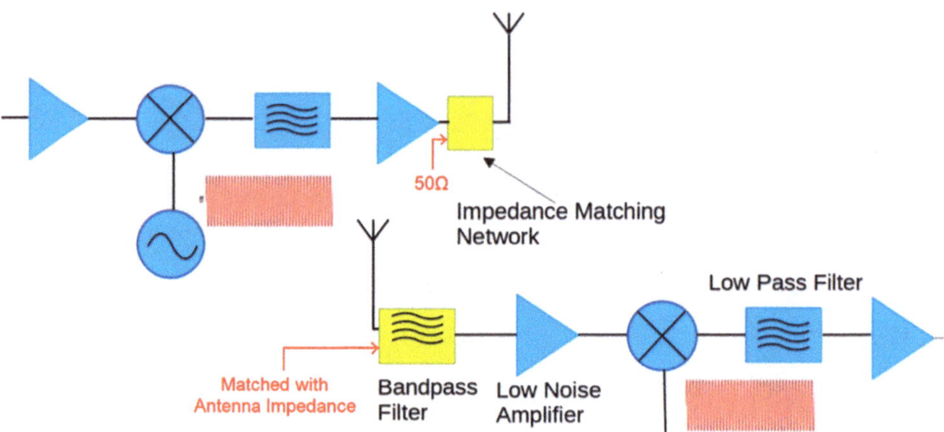

Fig. 10.8 The impedance matching for RF transmitter and receiver

the impedance of antenna is different from the 50 Ω so we use the impedance matching network before the antenna so that if we see from the input of the matching network point the impedance of the antenna is changed to 50 Ω using that impedance matching network, now for the rest of the devices in the circuit, most of the devices have their input and output impedance as 50 Ω but if some device does not have its impedance as 50 Ω, so with that device we also need to use the impedance matching network in order to match its impedance with the rest of the circuit. Now, in the receiver side we also have an antenna whose impedance is often different from 50 Ω but when we see the receiver circuit towards before the low noise amplifier point the impedance of our receiver circuit is 50 Ω so we need to match that impedance with the impedance of the receiver antenna so the bandpass filter that we are using, we also use it as the impedance matching network so that when we look from before the bandpass filter the impedance of the rest of the RF circuit is matched to the impedance of the receiver antenna and that impedance matching network at the transmit antenna ensures that the maximum power from transmitter is transferred to the antenna and then it is radiated in the wireless channel and also the impedance matching at the receive antenna ensures that the maximum power that is received by the antenna it is transferred to the receiving circuit as illustrated in Fig. 10.8.

10.9 Conclusion

This chapter has provided a thorough examination of impedance matching, a cornerstone concept for ensuring efficient and reliable operation in RF systems. We began by establishing the fundamental goal: to achieve maximum power transfer from a source to a load, which requires conjugate matching of their complex impedances, with 50 Ω serving as the standard reference in most RF circuits.

The consequences of failing to achieve this match were clearly detailed. An impedance mismatch creates reflected power, leading to reduced system efficiency and potential damage to components. We defined the key metrics used to quantify this mismatch: Return Loss (RL), which measures the suppression of reflected power in decibels, and the Voltage Standing Wave Ratio (VSWR), which describes the standing wave pattern on a transmission line. The critical relationship between VSWR and the percentage of power reflected was presented, demonstrating the severe performance degradation that occurs even with moderate VSWR values.

The chapter concluded with the practical solution: impedance matching networks. These networks, typically constructed from inductors and capacitors, are designed to transform a complex load impedance (such as that of an antenna) to appear as the desired system impedance when viewed from the source. Their indispensable role in both transmitter and receiver chains—ensuring that maximum power is delivered to the antenna for radiation and that maximum received signal is coupled into the receiver's front-end—was emphasized.

In summary, this chapter has equipped you with the principles to understand, quantify, and solve impedance matching challenges. Mastery of VSWR, return loss, and matching network design is essential for optimizing the performance of any RF link, forming a critical bridge between the theoretical performance of individual components like amplifiers and antennas and the practical efficiency of a complete communication system.

References

1. Tiwana M (2021) RF concepts, components and circuits for beginners. Udemy Inc., San Francisco, CA
2. Pozar DM (2011) Microwave engineering, 4th edn. John Wiley & Sons, Hoboken, NJ

The Scattering (S) Parameter 11

Contents

11.1 Introduction to the Network

Now, we discuss about the S-parameters which are used for the network analysis. So, what is a network? A network is a device that has one or more ports for example the network in Fig 11.1 has two ports which are input and output ports and each of these ports has two terminals.

The example of 2 port devices is amplifier and then you also have 1 port devices like antenna and you have 3 port devices for example the mixer is a 3-port device in which 2 ports are used as the input and 1 port is used as the output. Now, each of these ports can pass, reflect or absorb the power that is input to it [1].

11.2 The Network Analysis

Now, for the network analysis of a device we analyze all ports of a network device one by one, for example for the network device in Fig 11.2, if we want to analyze the port1, we would inject the RF power into that port and then we would measure the power that is

M. Pakdel, *Understanding RF Systems*, Synthesis Lectures on RF/Microwaves,
https://doi.org/10.1007/978-3-032-19227-1_11

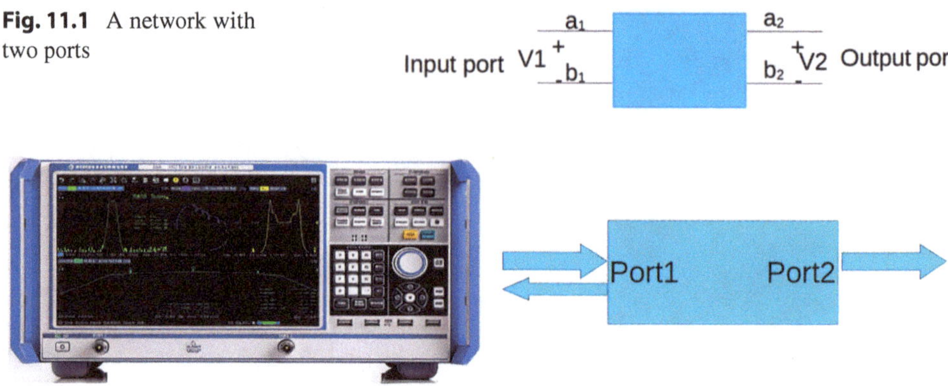

Fig. 11.1 A network with two ports

Fig. 11.2 A network device and the network analyzer

reflected by that port and also the power that is appearing at the output of the other ports and since the network response is mostly the frequency dependent so we analyze each port of the device over range of frequencies and this analysis is done using a device which is called as the network analyzer [2].

11.3 The S Parameters

Now, in order to characterize the complete network behavior of a device, we use the S parameters which is also known as the scattering parameters and also we call them as the high frequency parameters and they are calculated based upon the power measurements on each port and we calculate these parameters using the power measurements because at high frequencies it is difficult to measure the voltage and currents at each port and the S parameter is represented by symbol S_{xy}, where y represents the port in which the power is injected and x represents the port from which the power is emerging. For example, if we want to calculate the parameter S_{xy} we would use the following equation.

$$S_{xy} = \frac{\textit{The power leaving port } x}{\textit{The power entering port } y} \tag{11.1}$$

11.4 The Common Parameters

Now, we will discuss what the common S parameters and what their common names are. First, we would calculate the S_{11}, and a 2-port device is shown in Fig. 11.3.

As you can see in Fig. 11.3, the input port is a_1b_1, and the output port is a_2b_2 and the input port has 2 terminals, in the terminal a_1 the power is injected and from the terminal b_1, the power leaves that port, similarly in the case of second output port the power is injected

Fig. 11.3 A 2-port device

in the terminal a_2 and it leaves from terminal b_2. Now, in order to calculate the S_{11}, we inject the power in port 1through the terminal a_1 and we measure the power that is leaving the port b_1 and when we divide b_1 by a_1 we get the parameter S_{11} and that parameter is related to the input matching and that means it is related to the return loss, VSWR, and so that parameter should be as small as possible and it should be equal or less than −15 dB ($S_{11} \leq -15$ dB) for good operation. Now, when we inject power at terminal a_1 and measure the power that is leaving the terminal b_2 and divide b_2 by a_1 then we get the gain of the amplifier (device). So, how much the power is entering the input port and how much the power is leaving the output port if we divide the power at the output port by the power at the input port so we get the gain of the amplifier which should be as higher as possible and then we have the parameter S_{12} for example if we inject the power at port a_2 and measure the power that is coming out of the port b_1 so in that case it is undesirable power which is flowing in the wrong direction from the output to the input, so that means that parameter indicates the reverse isolation and that parameter should be as small as possible and then we have the parameter S_{22} and when the power is injected at the port a_2 and we measure the power that leaves from port b_2 and we divide b_2 by a_2 then we get the parameter S_{22} which is related to the output matching of the output port with the load and that parameter should be as small as possible and it should be small than or equal to -15 dB ($S_{22} \leq -15$ dB). Now, these S parameters have complex values and those parameters can be represented as N × N matrix where N indicates the number of ports in a network device, for example if we have a 1-port device then in that case the matrix will have a 1 × 1 dimension and that means it would only have one S parameter which is the S_{11} ($[S_{11}]$) and if we have a 2-port device, in that case we can represent the S parameters as the 2 × 2 matrix as below.

$$The\, S\; parameters\; matrix\; for\, a\, 2-port\; device = \begin{bmatrix} S_{11} & S_{12} \\ S_{21} & S_{22} \end{bmatrix} \tag{11.2}$$

Also, when these network devices are connected in series, in that case we can cascade those S parameters in order to find the overall system response.

11.5 Conclusion

This chapter has provided a foundational introduction to Scattering Parameters (S-parameters), establishing them as the essential analytical tool for characterizing RF and microwave components at high frequencies. We began by framing components as multi-port networks and outlined the methodology of network analysis, which uses a Vector

Network Analyzer (VNA) to systematically measure how a device reflects, transmits, and absorbs power across its ports.

The core of the chapter defined the S-parameter formalism, where each parameter S_{xy} represents the ratio of a power wave leaving port x to the power wave incident at port y. We detailed the practical significance of the key parameters for a two-port device: S_{11} and S_{22} as measures of input and output impedance matching (directly linked to VSWR and return loss), S_{21} as the forward gain or transmission, and S_{12} as the reverse isolation. The chapter emphasized the performance criteria for these parameters, such as requiring $S_{11} \leq -15$ dB for good impedance matching.

Finally, the concept of organizing these complex values into an S-parameter matrix was introduced, providing a compact and powerful mathematical representation of a network's complete behavior. This matrix formalism enables the analysis of multi-port devices and, crucially, allows for the cascading of individual component responses to model and predict the performance of entire RF subsystems.

In summary, this chapter has equipped you with the language and concepts to interpret S-parameters, which are indispensable for the specification, design, simulation, and validation of modern high-frequency circuits. Mastery of this topic forms a critical bridge between the theoretical operation of individual components like amplifiers and mixers and their measurable, real-world performance in interconnected systems.

References

1. Tiwana M (2021) RF concepts, Components and circuits for beginners. Udemy Inc., San Francisco, CA
2. Pozar DM (2011) Microwave engineering, 4th edn. Wiley, Hoboken, NJ

The Smith Chart for Impedance Matching 12

Contents

12.1 Introduction to the Smith Chart

Now, we would discuss about the fundamentals of Smith chart and we are also going to discuss one of the important applications of Smith chart which is in the impedance matching and we know that the impedance matching networks are important for example when we want to match the transmitter circuit with the antenna so that the maximum power is transferred to the antenna and then that power is radiated over the wireless channel as demonstrated in Fig. 10.8. So, these impedance matching networks can be designed using the Smith chart. The Smith chart has many different applications and the inventor of Smith chart was Philip H. Smith wrote a 200-page book about the applications of the Smith chart. The most important application of Smith chart is in the impedance matching and designing of the impedance matching networks. The Smith chart provides us with the graphical method of solving RF problems with minimum calculations. Now, with the advent of

M. Pakdel, *Understanding RF Systems*, Synthesis Lectures on RF/Microwaves,
https://doi.org/10.1007/978-3-032-19227-1_12

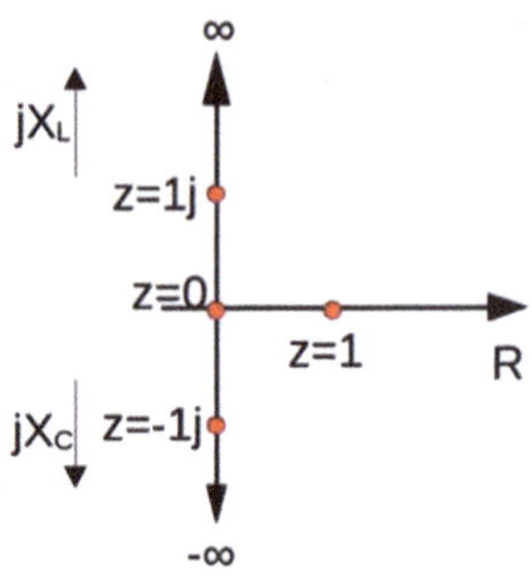

Fig. 12.1 Plotting the impedances on the cartesian coordinates

digital computers, the Smith chart is still used in order to visualize impedances and the matching networks [1].

12.2 The Basics of Smith Chart

Now, the first question that may arise in your mind that why we cannot use the cartesian coordinates in order to plot the impedances on the cartesian coordinates? For example if we have a purely resistive impedance with the value of 1, we can plot it along the X axis which is the resistance axis and if we have a case of short circuit with zero impedance we can plot it in the origin of cartesian coordinates and if we have a purely inductive impedance with the value of 1 we can plot it along the jX_L axis or along the positive reactance axis and if we have a purely capacitive load with the value of −1 we can plot it along the negative reactive axis ($-jX_C$ axis) as shown in Fig. 12.1.

Now, the problem with using the cartesian coordinates is that since the resistance can never be negative so we are only using one side of the cartesian coordinates and secondly for the case of open circuit we have infinite impedance and it is difficult to plot the infinite impedance along the reactive axis so what we do is that we use the Smith chart in the case of Smith chart the resistive line (R axis) is the horizontal line that is passing through the center of the Smith chart as depicted in Fig. 12.2.

While the positive reactance axis or the inductive axis is traversed in the upper semicircle, similarly the negative reactance axis is also traversed in the lower semicircle, so the origin-point in the Cartesian coordinates which indicates the impedance of zero or the short circuit is represented with a point as demonstrated in Fig. 12.2. Also, the impedance point of Z = 1j which is purely inductive impedance it corresponds to the point of pure inductive load in Fig. 12.2 and the impedance point of Z = −1j which is purely capacitive impedance it corresponds to the point of pure capacitive load in Fig. 12.2, while the open circuit point that lies in the infinity in Fig. 12.1 it corresponds to the point of open circuit in Fig. 12.2 and the point with impedance value of 1 (Z = 1) in Fig. 12.1, which is purely resistive impedance it corresponds to the point of Z0 system impedance in Fig. 12.2. So,

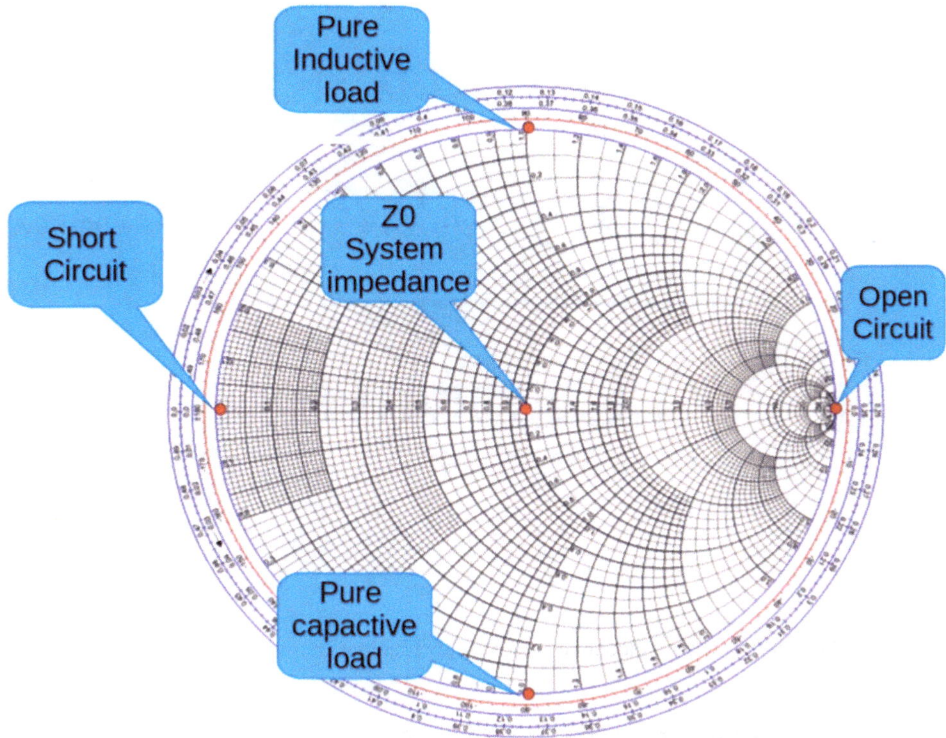

Fig. 12.2 The Smith Chart

in summary, the horizontal line in Fig. 12.3, indicates the purely resistive line along which the inductance value is zero.

Also, the upper part is used to indicate the inductive region that has a positive reactance and that lower part is used to indicate the capacitive region that has negative reactance. Now, the circles on the Smith chart have a constant resistance value for example in Fig. 12.4 the middle circle has the constant resistance value of 1, and the bigger diameter circle has the constant resistance value of 0.4, and the lower diameter circle has the constant resistance value of 3, so as we move along the circle the reactance at each point may change but the resistance value is kept constant for example in the bigger diameter circle, the resistance value is kept constant at 0.4 as illustrated in Fig. 12.4.

Then we have the constant reactance curve for example the first curve or arc at the up right of Fig. 12.4, it has the inductance value of 3 or it has the reactance value of 3 (3j), and the next curve or arc has the reactance value of 2 (2j), and if you see the next arc it has the reactance value of 1.6 (1.6j), now as you move along the arc the value of resistance may change but the value of inductance is kept constant as shown in Fig. 12.5 [2].

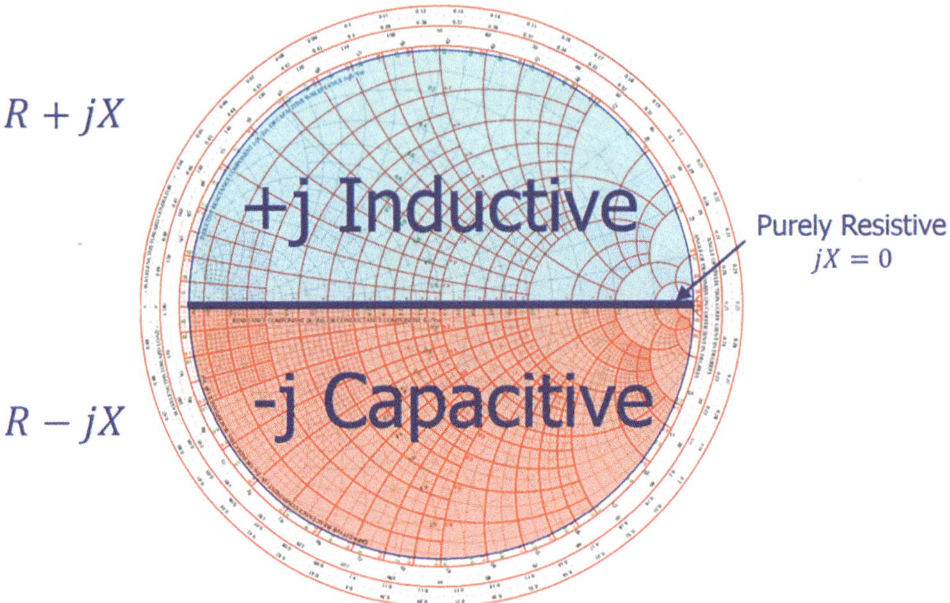

Fig. 12.3 The purely resistive, inductive and capacitive regions in Smith chart

12.3 The Complex Impedance on Smith Chart

Now, how do we plot the complex impedances on the Smith chart? First, we take the example of system impedance that is standardized to 50 Ω in most of the cases so the system impedance Z_0 is equal to 50 Ω and we would normalize that impedance by dividing it with 50 so we get the normalized system impedance as 1 ($Z_0 = 1$) so we plot it on the resistance line with the value of 1 since there is no reactance in that impedance and if we have a load impedance $Z_L = 25 + 40j$, first we will normalize it by dividing it by 50 so the normalized impedance that we get is $Z_L = 0.5 + 0.8j$, and now the 0.5 is the resistance part and the 0.8 is the reactance part so we will find the resistance circle that has the value of 0.5 and then we will find the reactance curve that has the value of 0.8 so on that curve we will find its intersection with the resistance circle and that intersection point is the load impedance that has now being plotted on the Smith chart as depicted in Fig. 12.6.

Now, if we add an inductor in series with the load impedance in that case the inductive reactance of the inductor would get added to the reactance of the impedance and as a result the overall load impedance would move in the clockwise direction on the constant resistance circle and if we add a capacitor in series with the load impedance, in that case the reactance of the capacitor would get subtracted from the reactance of the load impedance and as a result that load impedance would move in the counter clockwise direction on the constant resistance circle as illustrated in Fig. 12.7.

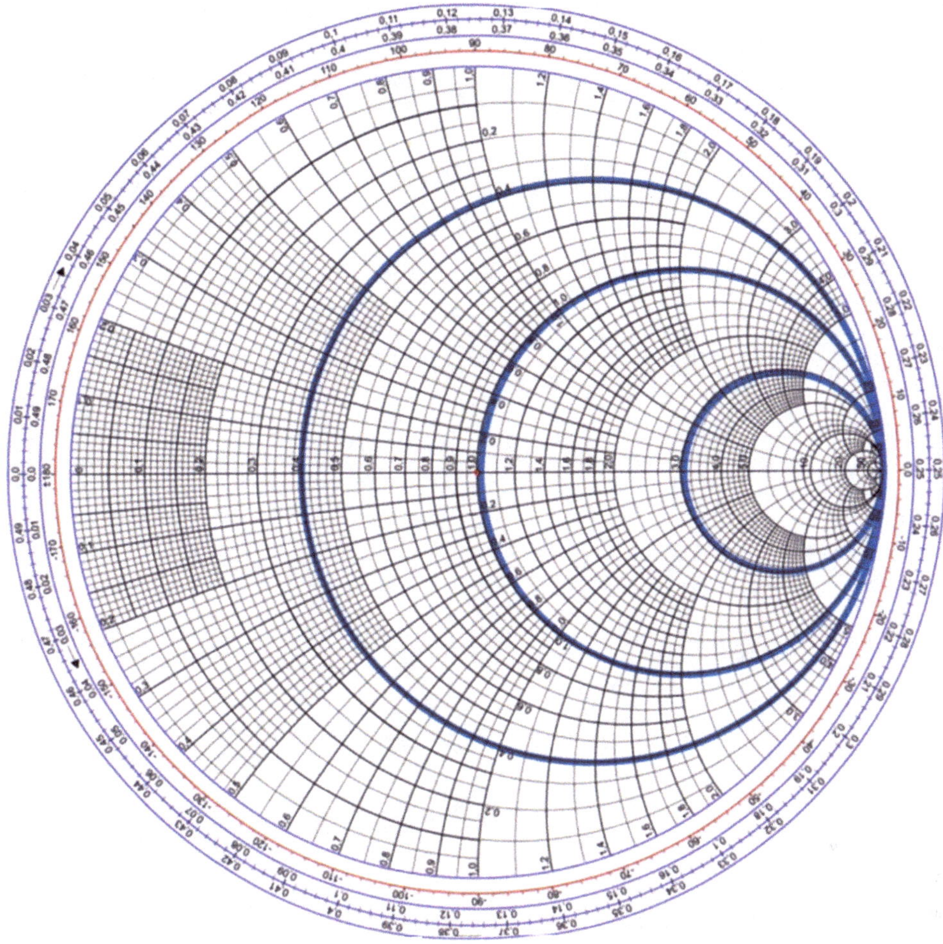

Fig. 12.4 The constant resistance circles

12.4 Converting the Impedance to Admittance

Now, we know that the admittance is the inverse of impedance and using Smith chart we easily convert the impedance to the admittance, for example if we have an impedance point on the Smith chart and we want to convert it to the admittance so we will use the compass and we will take the center of Smith chart as the center point and we will use a compass to draw a circle around the center point so that the impedance lies in the circle, now opposite to that impedance we have the admittance value on the circle so in that way we convert the impedance value into the admittance value as shown in Fig. 12.8.

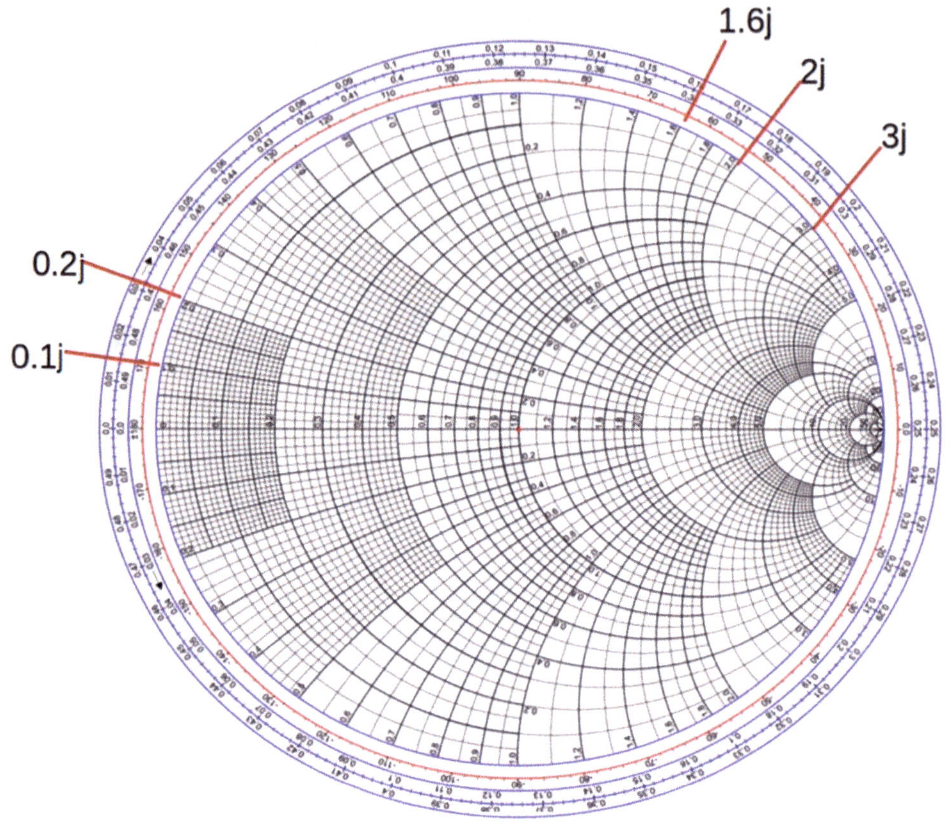

Fig. 12.5 The constant reactance curves

12.5 The Combined Impedance-Admittance Smith Chart

We can also convert an impedance Smith chart to an admittance Smith chart by just flipping the impedance Smith chart for example in Fig. 12.9 we have the impedance Smith chart and when we flip it becomes the admittance Smith chart.

Now, we combine the impedance and the admittance Smith charts into one Smith chart and it is depicted in Fig. 12.10.

Now, when we convert the impedance to the admittance, the resistance circles get converted into the conductance circles, since the conductance is the inverse of the resistance as the following equation.

$$G = \frac{1}{R} \tag{12.1}$$

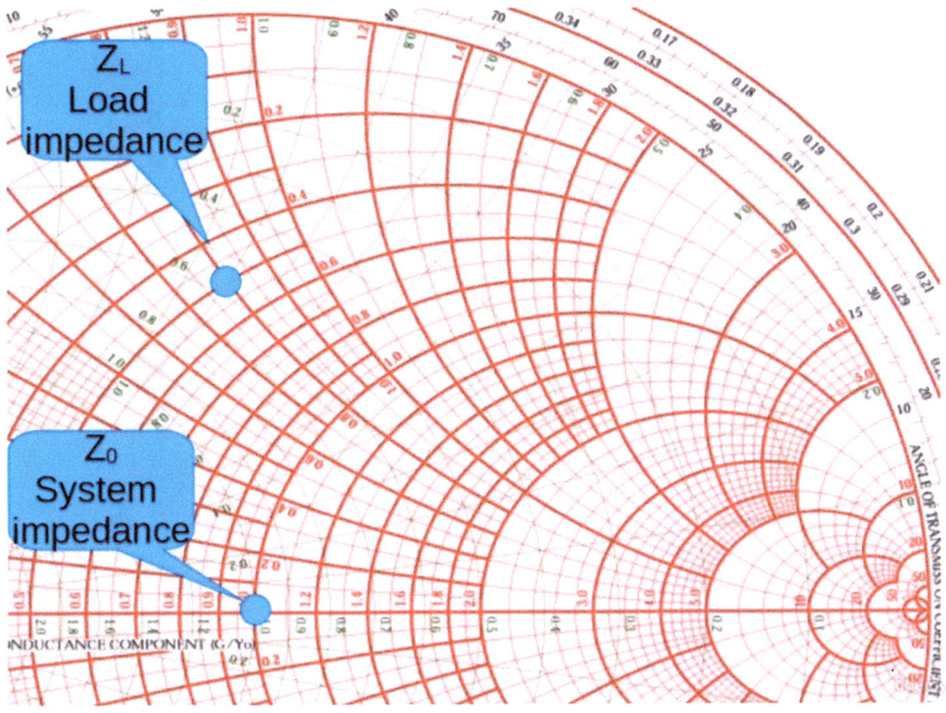

Fig. 12.6 Plotting the complex impedance on the Smith chart

So, you can see the resistance and conductance circles in Fig. 12.10. Similarly, the reactance curves get converted into the susceptance curves since the susceptance is the inverse of the reactance as the equation below.

$$B = \frac{1}{X} \tag{12.2}$$

Also, you can see the reactance and susceptance curves in Fig. 12.10. In overall, the relation between admittance and impedance is given by the following equation.

$$Y = \frac{1}{Z} \tag{12.3}$$

12.6 Adding Reactance to the Load Impedance

Now, before we do the impedance matching using the Smith chart, we need to understand what happens when we add reactive element either in series or parallel to the load impedance and how the position of overall the impedance changes on the Smith chart. For example, for the left arrow in Fig. 12.11, if we add a capacitor in series to the load impedance,

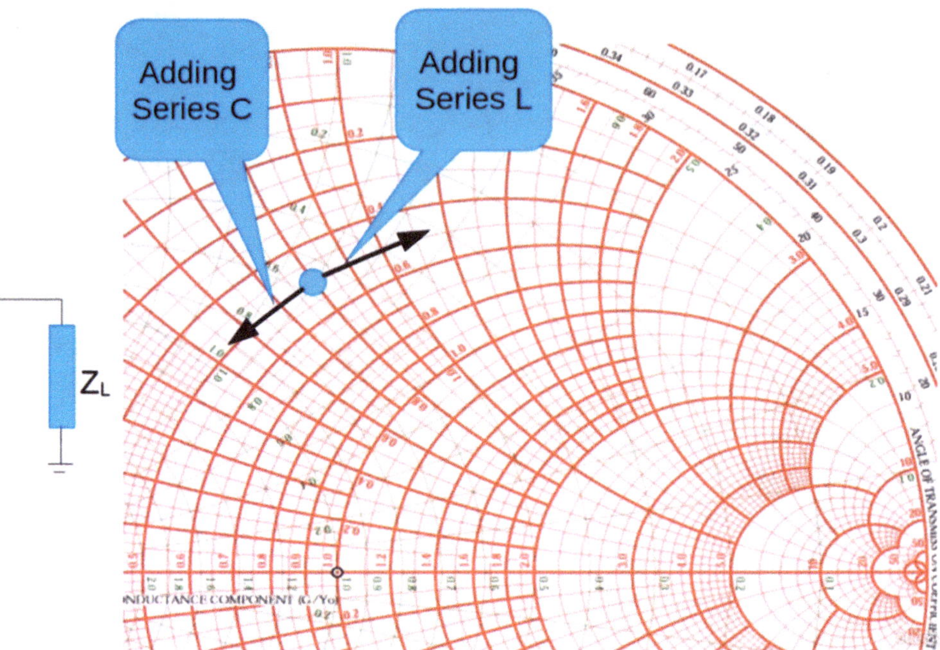

Fig. 12.7 Adding series inductor or capacitor to a load

in that case the overall load impedance would move in the counterclockwise direction on the constant resistance circle. Similarly, if we add an inductor in series to the load impedance, then adding that inductance would move the overall load impedance in the clockwise direction on the constant resistance circle as demonstrated by the second leftmost arrow in Fig. 12.11. On the other hand, if we add a capacitor in parallel with the load impedance, in that case the load impedance would move in the clockwise direction on the constant conductance circle as you can see in the second rightmost arrow in Fig. 12.11. Now, if we add an inductor in parallel with the load impedance in that case that load impedance would move in the counterclockwise direction on the constant conductance circle as demonstrated by the right arrow in Fig. 12.11.

12.7 The L-Networks for Impedance Matching

The L-networks are a type of networks that are used in the impedance matching and those L-networks are further subdivided into 8 different types and depending upon the position of the load impedance on the Smith chart we choose an L-network, for example if the load impedance lies in the top left Smith chart in Fig. 12.12 which is closed by a tick curve line, then we choose the type 1 L-network in which we have an inductor in series and then we have a capacitor in parallel, and if the load impedance lies in the region which is closed by

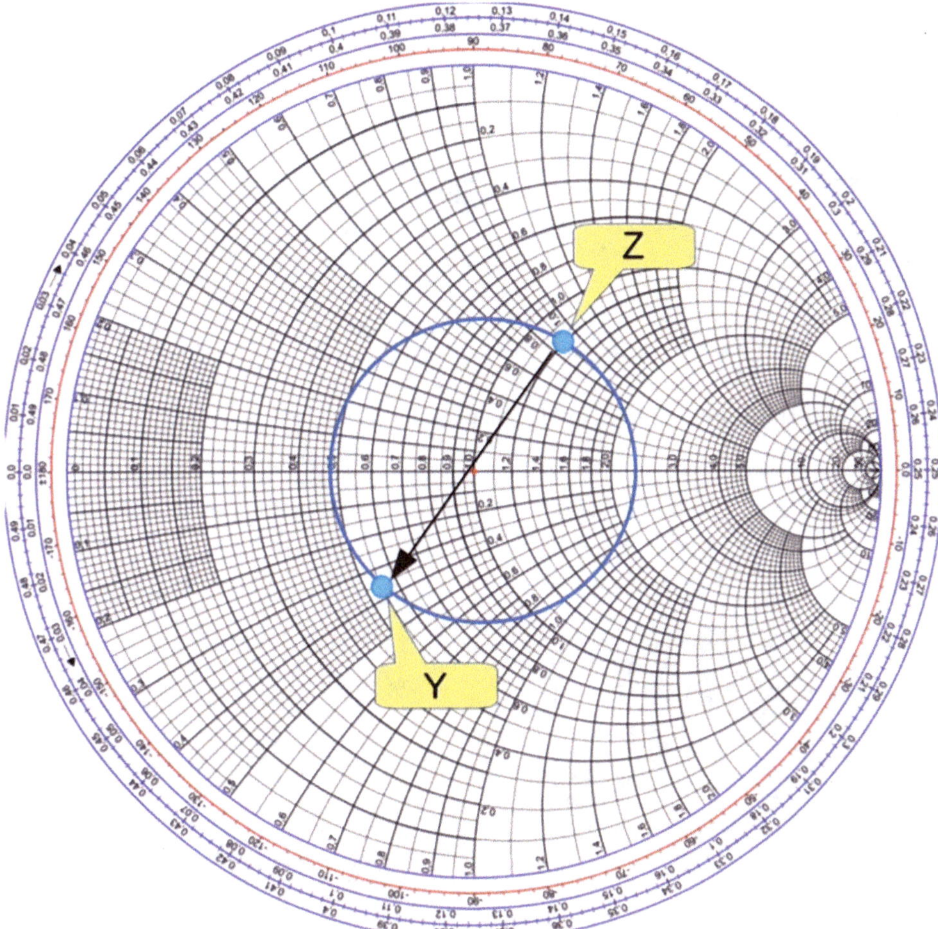

Fig. 12.8 Converting the impedance to the admittance

a tick curve line in the middle top Smith chart in Fig. 12.12, then we choose the type 2 L-network where first we have a capacitor in series and then we have an inductor in the parallel, and if the load impedance lies in the region which is closed by a tick curve line in the left bottom Smith chart then we can choose the type 3 L-network that has first a capacitor in parallel then an inductor in series and if the load impedance lies in the region which is closed by a tick curve line in the middle bottom Smith chart then we can choose the type 4 L-network that has an inductor in parallel first and then a capacitor in series. Now, if the load impedance lies in the region which is closed by a tick curve line in the top right Smith chart then either we can choose the type 5 or the type 7 L-network and those networks consists of two capacitors one in series and one in parallel and we can have the parallel capacitor first or the series capacitor first and then if the load impedance lies in the region which is closed by a tick curve line in the bottom right Smith chart then either we can

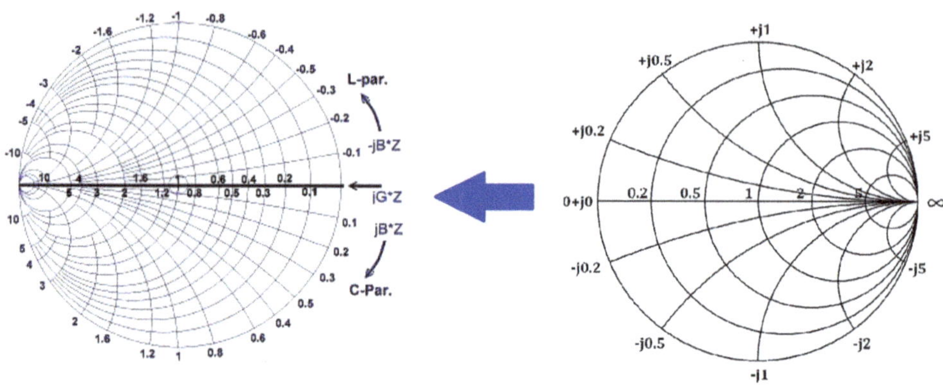

Fig. 12.9 Converting the impedance to admittance Smith chart

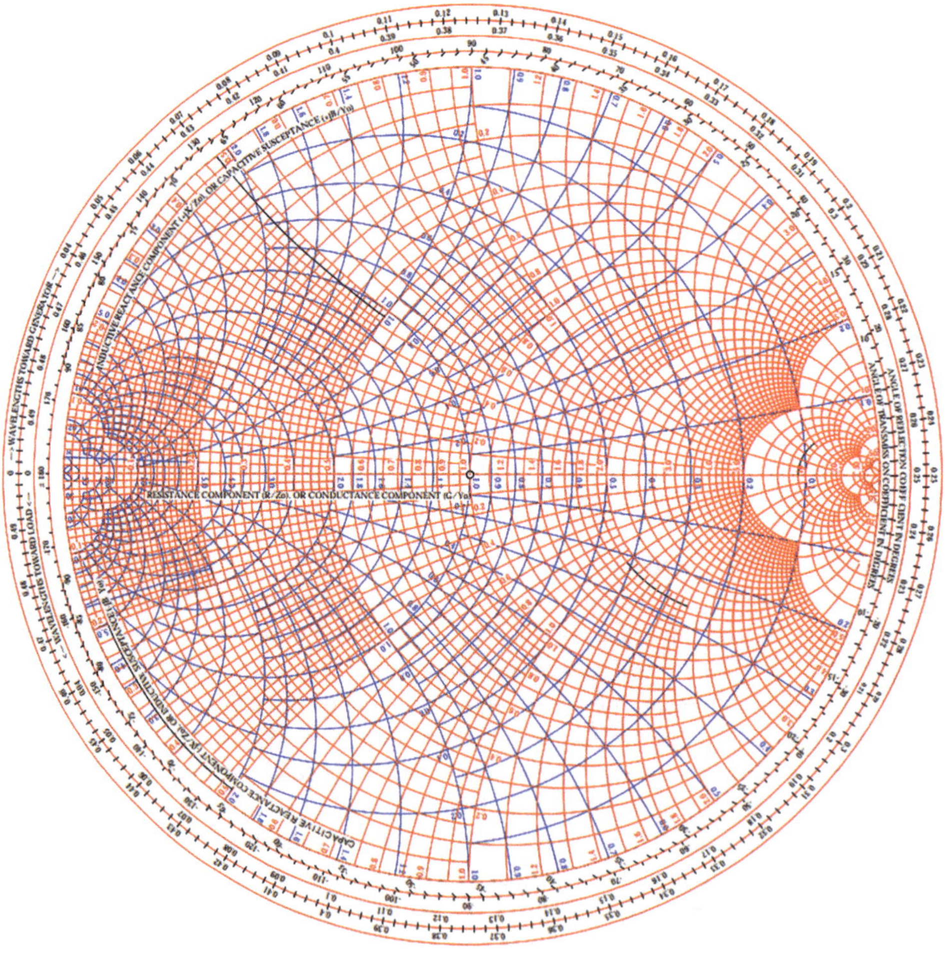

Fig. 12.10 Combining the impedance and the admittance Smith charts

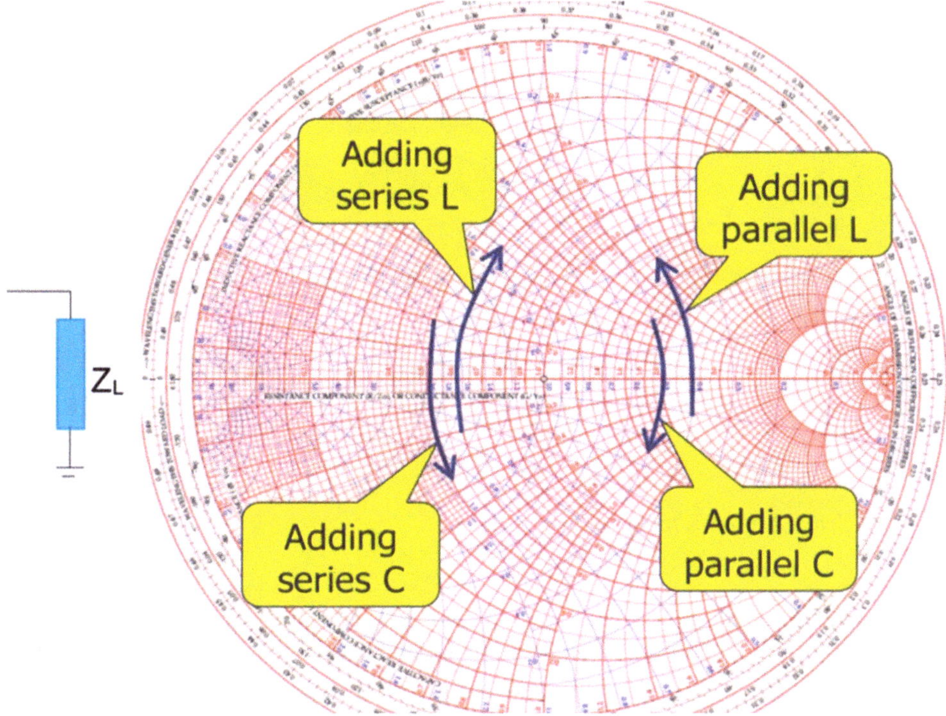

Fig. 12.11 Adding reactive elements to the load impedance

choose the type 6 or the type 8 L-network and those networks consists of two inductors one in series and one in parallel and we can have the parallel inductor first or the series inductor first as demonstrated in Fig. 12.12.

So, it is clear that depending upon the position of the load impedance we may have the possibility to choose from more than one L-network types.

12.8 The Impedance Matching Using Smith Chart

Now, we do an example of impedance matching using the Smith chart and we have a load impedance of $Z_L = 75—60j$, so first we normalize the load impedance by dividing it by 50 and so we get the normalized load impedance of $Z_L = 1.5—1.2j$, so that load impedance would lie at the intersection of 1.5 resistance circle and – 1.2j reactance curve and so we get the intersection of those two and we get the normalized load impedance plotted on the Smith chart as illustrated in Fig. 12.13.

Now, the standard impedance is also demonstrated in Fig. 12.13 at the resistance circle of 1, so we have to match that load impedance with the standard impedance and that means we need to move that point in such a way on the Smith chart that it reaches there at the

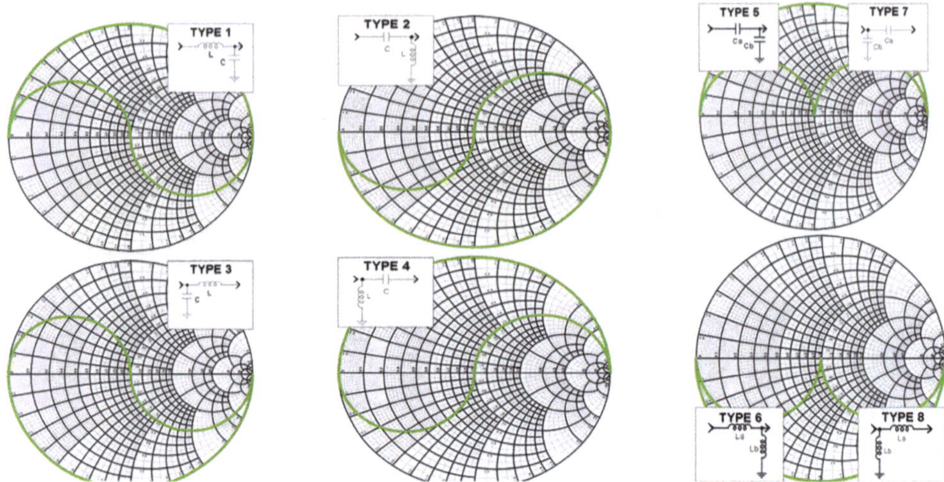

Fig. 12.12 The L-networks for impedance matching

standard load impedance. Now, since we can see in Fig. 12.13 that the load impedance lies in that region, so we choose the type 2 L-network that has a capacitor in series and an inductor in parallel. Now, as demonstrated in Fig. 12.14, we are on the conductance circle and we have to move to the point at standard resistance circle of 1, so first we will go counterclockwise on that conductance circle because going counterclockwise on the conductance circle means that we are adding an inductor in the parallel. At the load the susceptance value is 0.32j and in the intersection point of the resistance circle of 1 and the constant conductance circle we can see that the susceptance is −0.5j as shown in Fig. 12.14.

So, the required susceptance is −0.82j (−0.5j—0.32j) and that is the normalized susceptance so we will take the inverse of the normalized susceptance in order to calculate the normalized inductive reactance (−1/0.82j = 1.22j), so the normalized inductive reactance is 1.22 that we get and then we do multiply it with 50 to get the actual inductive reactance (1.22 × 50 = 61), now we know that the inductive reactance is defined by the following equation.

$$X_L = 2\pi fL \tag{12.4}$$

So, we can easily calculate the value of L from the Eq. (12.4), so we divide the inductive reactance value which is 61 by 2πf and we get the inductance value and then we add that inductance value in parallel as demonstrated in Fig. 12.14. Now, once we are at the intersection point of the resistance circle of 1 and the constant conductance circle, we can easily move from that point to the standard point in the constant resistance of 1 circle which is our target impedance value as depicted in Fig. 12.15 and we move −1.2j units in the counterclockwise direction on the constant resistance circle and that means we are adding a capacitor in series which has a normalized capacitive reactance of 1.2 so we multiply it

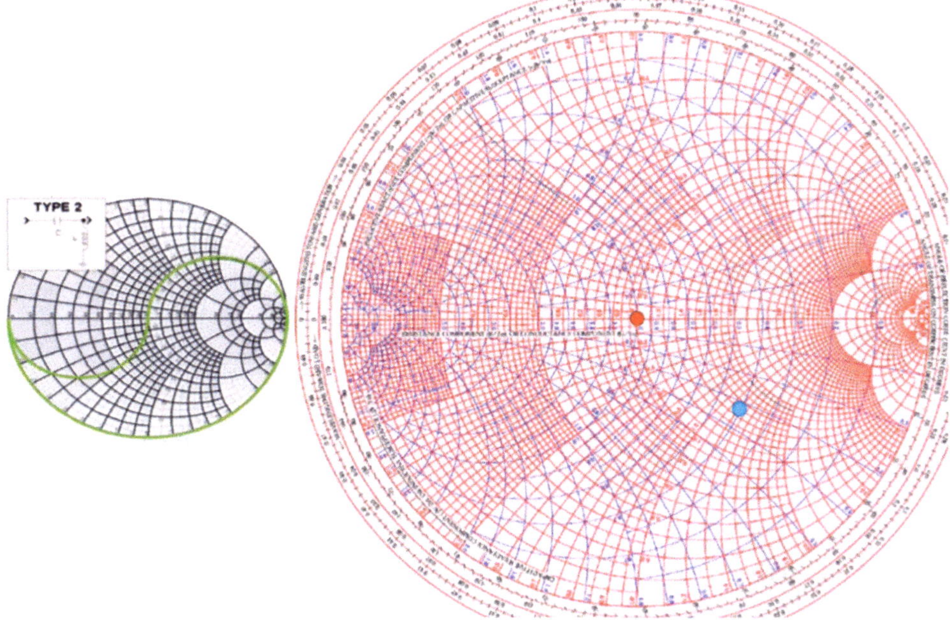

Fig. 12.13 The impedance matching example

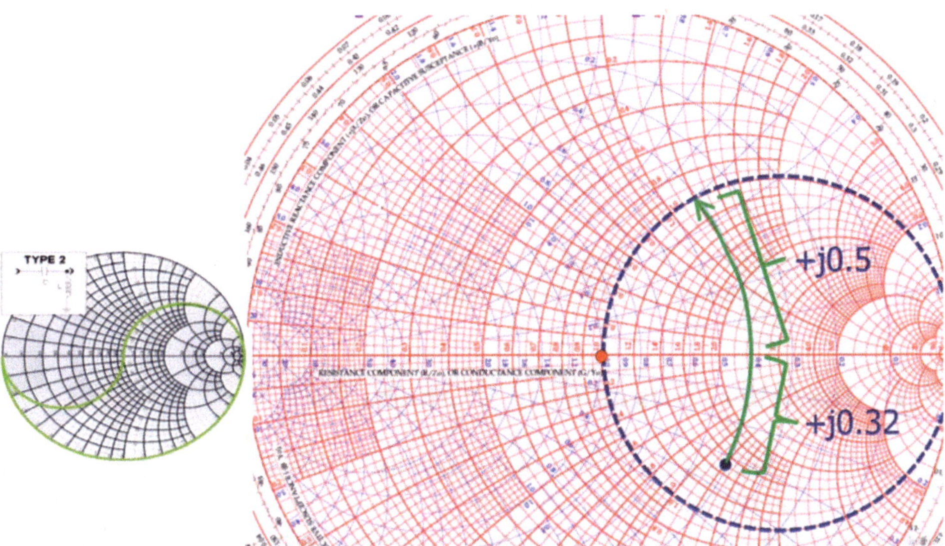

Fig. 12.14 Calculating the required inductive susceptance

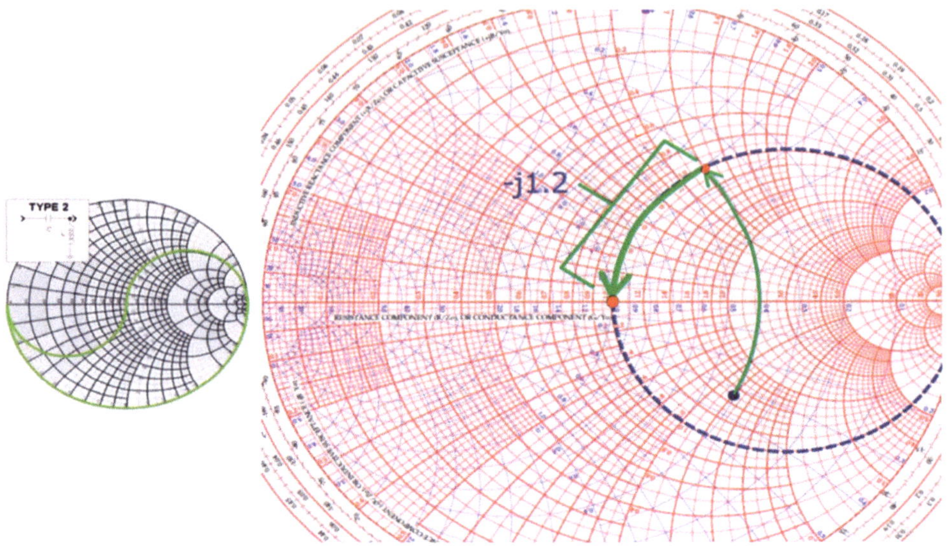

Fig. 12.15 Calculating the required capacitive reactance

with 50 to get the actual capacitive reactance value and we know that the capacitive reactance is given by the following equation.

$$X_C = \frac{1}{2\pi fC} \tag{12.5}$$

So, we can easily calculate the C using the equation below.

$$C = \frac{1}{X_C \times 2\pi f} \tag{12.6}$$

Then, that is the capacitance that we need to add in series so now we have the inductance value and the capacitance value for that impedance matching network.

12.9 The Other Applications of Smith Chart

Now, the Smith chart has many other applications like it is used to plot and compute the complex reflection coefficient for the load and the source, it is also used to determine the length of transmission lines, the type of transmission lines that is used for load matching or balancing, the VSWR analysis and to determine and plot the input impedance.

12.10 Conclusion

This chapter has provided a comprehensive introduction to the Smith Chart, establishing it as an indispensable graphical tool for visualizing and solving impedance-related challenges in RF engineering. We began by framing the Smith Chart's primary application: the design of impedance matching networks, which are critical for maximizing power transfer and minimizing reflections in systems like transmitter-to-antenna interfaces.

The chapter built a foundational understanding of the chart's structure, explaining how it maps the infinite complex impedance plane onto a finite circle through constant resistance circles and constant reactance curves. This graphical representation solves the practical limitations of Cartesian coordinates for RF work. We detailed the process of plotting normalized complex impedances and demonstrated how the addition of series or parallel reactive components (inductors and capacitors) translates to predictable movements along these circles and curves.

A significant portion of the chapter was dedicated to the practical application of this theory. By introducing the combined impedance-admittance Smith Chart, we provided the framework for designing matching networks that combine both series and shunt elements. This led to a systematic exploration of the eight fundamental L-section matching network topologies, where the chart's regions dictate the correct choice of components. The chapter culminated in a detailed, step-by-step design example, walking through the process of transforming a given complex load impedance to the standard system impedance by calculating the required inductor and capacitor values directly from the chart's geometry.

In summary, this chapter has equipped you with the principles to use the Smith Chart not merely as a diagram but as a powerful visual calculator. Mastery of this tool is essential for efficiently designing matching networks, analyzing transmission lines, and understanding complex impedance transformations, forming a critical skill set for practical RF and microwave circuit design.

References

1. Tiwana M (2021) RF concepts, components and circuits for beginners. Udemy Inc., San Francisco, CA
2. Pozar DM (2011) Microwave engineering, 4th edn. John Wiley & Sons, Hoboken, NJ

The Low Noise Amplifier (LNA) 13

Contents

13.1 Introduction to the LNA

Now, we would discuss about the low noise amplifier (LNA). The LNA receives a very weak signal from the antenna and that is why that antenna must be matched with that LNA and to do the matching we use the bandpass filter and that bandpass filter has two purposes first it matches the impedance of the antenna with the impedance of LNA secondly it only allows the desired signal to pass through and it rejects the noise as shown in Fig. 13.1.

Now, even then the signal that is received at the input of the LNA it is in millivolts and the LNA amplifies that weak signal into volts, however if that LNA has even small noise it can degrade or it can compromise the input weak signal and so that LNA must have as low noise as possible of its own, now in order to understand this let us consider the scenario as depicted in Fig. 13.2.

As you can see in Fig. 13.2, the signal with the certain signal to noise ratio (S/N) is coming to the input of the LNA and the signal to noise ratio of the input signal can be calculated by the following equation.

$$\left(\frac{S}{N}\right)_{IN} = \frac{Signal\ Power}{Noise\ Power} \tag{13.1}$$

M. Pakdel, *Understanding RF Systems*, Synthesis Lectures on RF/Microwaves,
https://doi.org/10.1007/978-3-032-19227-1_13

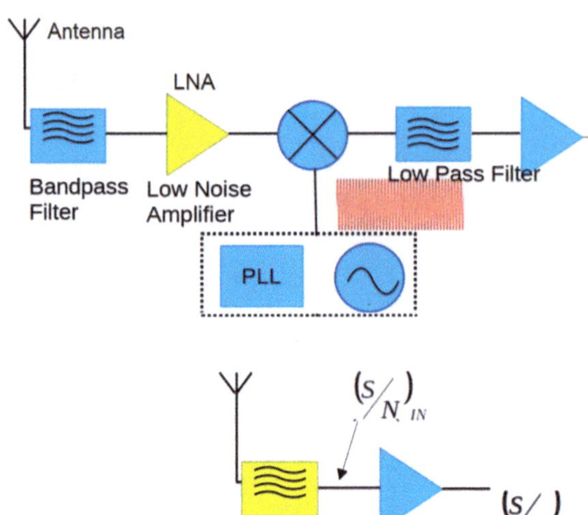

Fig. 13.1 The LNA in RF receiver circuit

Fig. 13.2 The signal to noise ratio at the input and output of LNA

Now, as the signal passes through the LNA then the signal to noise ratio at the output is given by the equation below.

$$\left(\frac{S}{N}\right)_{OUT} = \frac{Gain \times Signal\ Power}{Gain \times Noise\ Power + LNA\ Noise} \tag{13.2}$$

We can see in Eq. (13.2) that the signal power has been amplified by the gain of LNA but at the same time the noise power is also being amplified by the gain of LNA and we can also see that the LNA noise has been added in the equation, so we can see that the signal to noise ratio of the output is degraded because of that LNA noise, now if that LNA noise is zero in that case the signal to noise ratio of the input would be equal to the signal to noise ratio of the output and then that means the LNA is not introducing any degradation into the input signal, however we know that LNA will have some noise so the signal to noise ratio at the output of LNA degrades according to the noise of the LNA and smaller that LNA noise is smaller the degradation of the signal to noise ratio at the output [1].

13.2 The Noise Figure (NF)

Now, we would discuss about another very important parameter of the RF devices which is the noise figure (NF). The noise figure (NF) of a device for example the LNA as illustrated in Fig. 13.3 is defined by the following equation.

$$NF = \frac{(S/N)_{IN}}{(S/N)_{OUT}} \tag{13.3}$$

Fig. 13.3 The NF definition for LNA

The NF is used to measure that how much the device degrades the input signal to noise ratio ($(S/N)_{IN}$) and if the device does not degrade the input signal to noise ratio in that case the signal to noise ratio at the input and the signal to noise ratio at the output are the same, so the value of NF is 1. So, for the lossless devices, the NF value is 1 but if that device has some noise in that case the signal to noise ratio at the output is lower than the signal to noise ratio at the input, so the NF value is greater than 1 (NF > 1). Now, the NF is a very important parameter as far as the LNA are concerned, because the input signal to the LNA is a weak signal and the LNA must not degrade the weak signal so the NF of LNA must be as closed to 1 as possible. Normally the NF is quoted in dBs so if we want to convert that NF into dBs, we use the following equation.

$$NF(dB) = 10\log_{10}(NF), \qquad NF \geq 0dB \tag{13.4}$$

So, for the NF value of 1 that NF value in dB would be 0 dB and that means for the LNA the NF must be as close to 0 dB as possible [2].

13.3 The Characteristics of LNA

Now, the other important parameters of LNA are what is the frequency response or bandwidth that the LNA can amplify and with what gain it can amplify those frequencies and what is the output power of the LNA, what is the power efficiency of the LNA, what is the input and the output impedance of the LNA and how much that LNA is linear and that linearity is measured in terms of 1 dB compression point and the third order intercept that we discussed earlier.

13.4 Conclusion

This chapter has provided a detailed examination of the Low Noise Amplifier (LNA), establishing its role as the most critical component in an RF receiver's front-end for determining overall system sensitivity. We began by defining the LNA's primary challenge: to provide significant power gain to a very weak signal from the antenna while introducing a minimal amount of its own internal noise. The degradation of signal quality was analyzed through the Signal-to-Noise Ratio (S/N), showing how a real amplifier reduces this ratio.

The chapter established Noise Figure (NF) as the paramount metric for this component. Defined as the ratio of the input S/N to the output S/N, a perfect noiseless amplifier has an NF of 1 (0 dB). For practical LNAs, the NF must be as close to this ideal value as possible, as it directly dictates the receiver's ability to detect faint signals. We detailed the calculation and specification of NF in both linear and logarithmic (dB) terms.

Finally, the chapter placed the noise figure within the broader context of LNA specifications. A complete performance profile includes its frequency response and bandwidth, input and output impedance for proper matching, output power, power efficiency, and crucially, its linearity as measured by the 1-dB Compression Point (P1dB) and Third-Order Intercept Point (IP3). These parameters collectively determine how an LNA interacts with the rest of the receiver chain.

In summary, this chapter has equipped you with the principles to understand why the LNA is not just another amplifier but a specialized device where noise performance is the primary design constraint. Mastery of NF and the associated specifications is essential for designing or selecting an LNA that maximizes the performance of the complete receiver system analyzed in previous chapters.

References

1. Tiwana M (2021) RF concepts, components and circuits for beginners. Udemy Inc., San Francisco, CA
2. Pozar DM (2011) Microwave engineering, 4th edn. John Wiley & Sons, Hoboken, NJ

The RF Attenuators and their Types 14

Contents

14.1 Introduction to the RF Attenuators

Now, we would discuss about the RF attenuators, the RF attenuators are normally represented by the variable and fixed symbols as shown in Fig. 14.1.

The purpose of the RF attenuator is to reduce the level of the input signal and pass it to the output and that reduction or attenuation in the level of the signal is defined in decibels or dBs, now these attenuators can be used at the input of the amplifier so that the amplifier is not operating in the saturation region because we know in the saturation region the amplifier introduces the nonlinearities into the signal and these attenuators can also be used at the input of the circuit to protect that circuit from high input signal and these attenuators also used to match the source impedance with the load impedance (impedance matching) [1].

M. Pakdel, *Understanding RF Systems*, Synthesis Lectures on RF/Microwaves,
https://doi.org/10.1007/978-3-032-19227-1_14

Fig. 14.1 The attenuator symbols

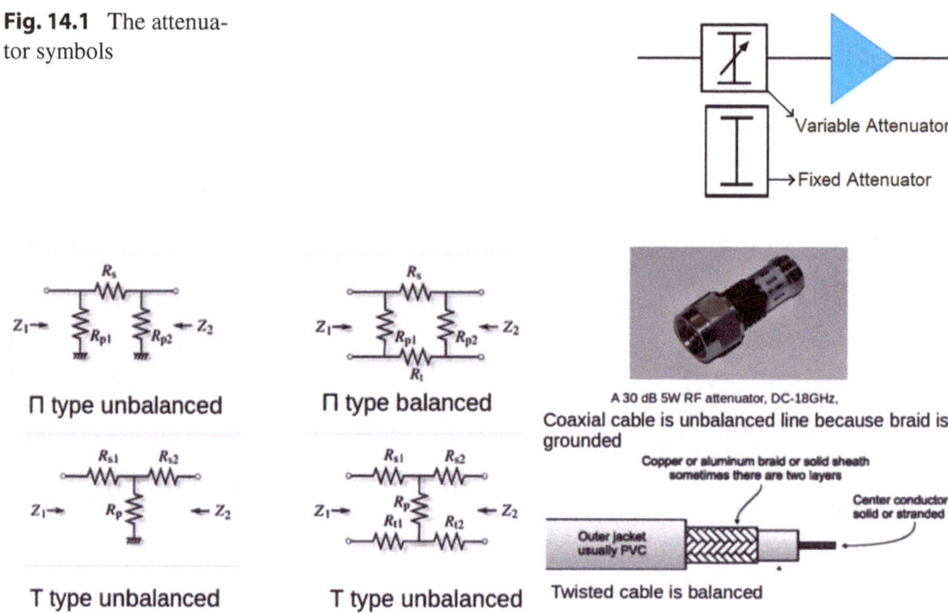

Fig. 14.2 The fixed type RF attenuators

14.2 The Fixed Type RF Attenuators

Now, we would discuss about the fixed type RF attenuator and normally we use the resistor network in order to make RF attenuators that have a fixed attenuation for example in Fig. 14.2, we have a resistor network that consists of 3 resistors that are connected in the shape of π and that π resistor network is unbalanced because that π network is grounded. However, if we add a resistor there and make the π network balanced so we get the resistor network which is π type and balanced and then we have the T type resistor network in which 3 resistors are connected in the form of a T and we can make it balanced those 2 resistors as demonstrated in Fig. 14.2.

Now, the unbalanced RF attenuators are used with the unbalanced load for example when we talk about the coaxial cable, in the coaxial cable you have a center conductor and then surrounding it you have the copper or the aluminum braid and that braid is grounded so the coaxial cable is unbalanced and we can use the unbalanced attenuators with that coaxial cable while when we talk about the twisted pair cable, it has two cables and it is balanced so we can use the balanced RF attenuators with the twisted pair cable. Now in Fig. 14.2, we have an example of 30 dB RF attenuator that has the maximum power rating of 5 W and it can operate from DC to 18GHz frequencies providing a fix attenuation value of 30 dB.

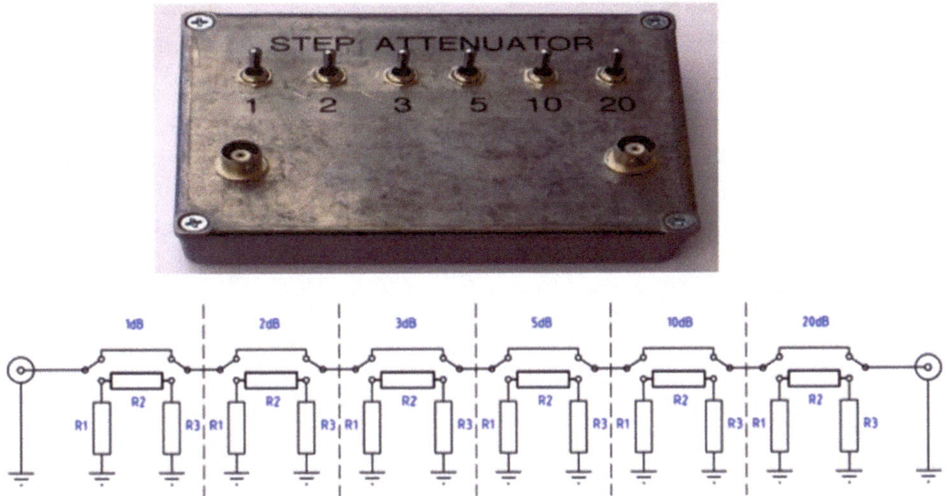

Fig. 14.3 The step type RF attenuator

14.3 The Step Type RF Attenuators

Now, we would discuss about the step type RF attenuators in which the level of attenuation can be changed in steps for example you can see in Fig. 14.3 an RF attenuator in which the attenuation can be changed in steps using those switches also you can view the internal circuit diagram of that RF attenuator in Fig. 14.3, now when we press the switch 1, the first stage unbalanced π type attenuator switches on the left in circuit diagram are move to another position and that resistor network is put into the circuit and that resistor network provides with an attenuation of 1 dB and if the switch 3 is pressed only then in that case the third stage unbalanced π type attenuator switches on the left in circuit diagram are move to another position and that resistor network is put into the circuit and that resistor network provides with an attenuation of 3 dB, an if we press switch 1 and switch 3 at the same time in that case those two resistor networks provide with a total attenuation of 1 dB plus 3dBs which is 4 dBs as depicted in Fig. 14.3.

Then you have the PIN diode based switched attenuators in which the PIN diodes are used as a switch, to switch different elements in or out of the circuit and to control the attenuation level and in such attenuators, we use digitally controlled select lines in order to select the level of attenuation as illustrated in Fig. 14.4.

As you can see in Fig. 14.4, we have 4 control lines that are being used in order to select the level of attenuation and the number of level of attenuations that are possible depend upon the control inputs for example if we have 4 input control lines then in that case we will have 2^4 or 16 levels of attenuation and if we have "n" input control lines then we will have 2^n level of possible attenuation values that can be digitally selected using with

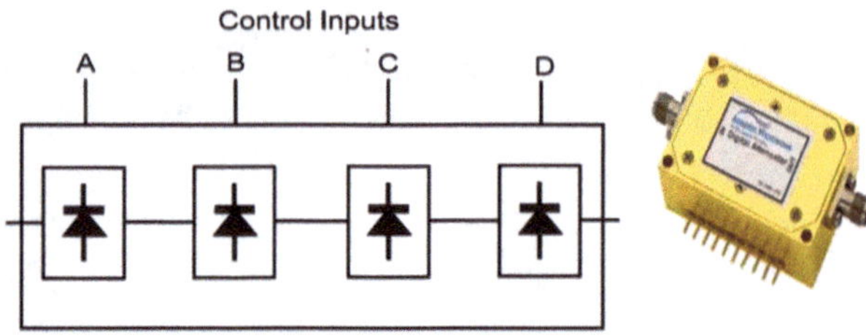

Fig. 14.4 The PIN diode based switched attenuators

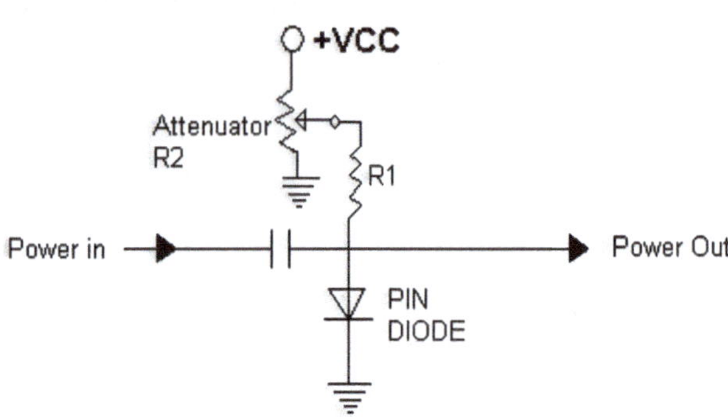

Fig. 14.5 An example of the variable RF attenuator

controlling select lines. Also, an example of PIN diode-based switched RF attenuator in which we have the input, the output and the select lines.

14.4 The Variable Type RF Attenuators

Now, we would discuss about the variable RF attenuators, the variable RF attenuators are used in the circuits where we continuously need to vary the attenuation level of a signal and the level of attenuation is controlled by using an analog voltage that is applied to the input control line, and an example of the variable RF attenuator is shown in Fig. 14.5.

As you can see in Fig. 14.5, that circuit is based on the PIN diode, the PIN diode has a particular property that in the forward bias as we increase the current through the PIN diode the resistance of the PIN diode decreases, now in that circuit we have the resistance R2 which is a potentiometer and by adjusting the potentiometer using that knob we can adjust the voltage that is appearing across the resistance R1 and also across the PIN diode and when the voltage across the diode is minimum in that case that PIN diode is reverse biased and that means the input signal (Power in) is going to the output without any opposition. However, as we increase the voltage using the R2 potentiometer so the more voltage appears across the PIN diode and that diode becomes forward biased and the input signal starts going to the ground through that PIN diode so as a result a less signal power is going to the output (Power out). Now, as we increase the voltage across the PIN diode using the R2 potentiometer, more current flows through the diode and the resistance of the diode decreases more and more signal then goes to ground through that PIN diode and less

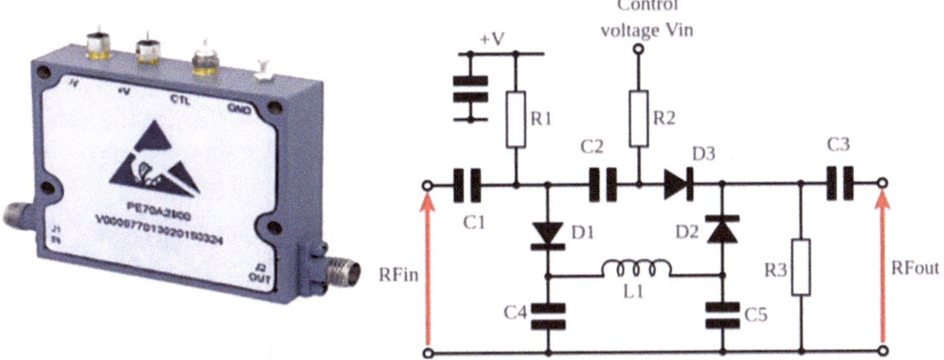

Fig. 14.6 The PIN diodes in bridge configuration

signal power reaches at the output so in this way the output signal power is attenuated. Now the PIN diodes are often used in the bridge configuration as depicted in Fig. 14.6.

As you can see in Fig. 14.6, the level of attenuation is being controlled by an analog voltage Vin we are applying there in the circuit diagram and the Vin has the minimum value and in that case the diode D3 would be reverse biased, however the diode D1 is forward biased so the input RF signal would be grounded through the diode D1 and it would not go to the output because the diode D2 is also reverse biased and so the maximum attenuation is achieved when the input voltage is minimum. Now, as we increase the voltage Vin in that case the diode D3 would be forward biased and the diode D1 would be reverse biased so in that case the input RF signal that we are applying it would start going to the output and if we further increase the voltage Vin in that case the current through the diode D3 would be further increased and its resistance would be further decreased and as a result more power of the input RF signal would start going to the output so in this way by increasing the Vin the attenuation of the circuit decreases. Now, the field effect transistors or the FETs can also be used to achieve the continuous variable level of attenuation and they act like a PIN diode so by using an analog control voltage we can control the level of attenuation that the FET is providing to the input RF signal and those FETs can also be used just like the PIN diode as a switch in order to switch different circuit elements in and out of the circuit and to achieve the step attenuation of the RF signal in that way. An example of a FET based variable attenuator which is the Avago AMMP-6640 chip and it can provide variable attenuation over a 20 dB range and that RF attenuator can work from DC to 40GHz range as illustrated in Fig. 14.7 [2].

14.5 The RF Attenuator Specifications

Now, we would discuss about the important RF attenuator specifications, the first one is the frequency range over which that attenuator operates, the second one is the attenuation type whether this attenuator provides the discrete or continuous attenuation and then we

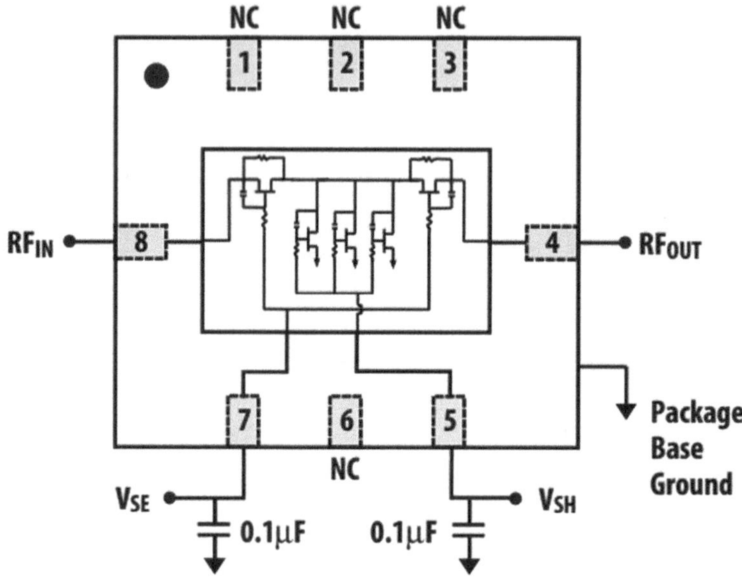

Fig. 14.7 An example of a FET based variable attenuator (Avago AMMP-6640)

have the attenuation resolution that means that what is the minimum attenuation step of that attenuator, it is 0.5 dB, 1 dB, etc. Then we have the attenuation range for example we have an attenuator that has 32 dB attenuation range and it has 6 control bits and corresponding to 6 control bits we have 64 levels of attenuation and that means that attenuation step is of 0.5 dB, so the minimum attenuation step is of 0.5 dB. So this attenuator can provide us the attenuation from 0 to 31.5 dB in the steps of 0.5 dB and then you have the accuracy of the attenuator that is reported across the frequency range and then you have the insertion loss of the attenuator, the insertion loss is the attenuation of the attenuator when we select the 0 dB as the attenuation of the attenuator and we apply that attenuator in the circuit, then in such condition the attenuation that is provided by this attenuator is called as the insertion loss, so that is the loss of a device itself when we have not applied any attenuation to this device. Then you have the power handling capability of the attenuator and that is what is the maximum input power that the attenuator can handle without breaking down and then you have the input impedance of the attenuator and finally you have the linearity of the attenuator that is measured as the third order intercept and this value is normally reported across the attenuation range.

14.6 Conclusion

This chapter has provided a comprehensive examination of Radio Frequency (RF) attenuators, essential passive components for precise signal level management. We began by establishing their core function: to reduce signal power by a calibrated amount, measured in decibels (dB), for critical applications such as preventing amplifier saturation, circuit protection, and impedance matching.

The chapter systematically classified attenuators by their control mechanism. First, fixed attenuators were analyzed, detailing their implementation through resistive pi (π) and tee (T) networks in both balanced and unbalanced configurations for different transmission line types. The discussion then progressed to step (switched) attenuators, which offer discrete, selectable attenuation levels, implemented through mechanical switches or rapidly acting PIN diodes controlled by digital logic lines.

A significant focus was placed on continuously variable attenuators, which allow for analog control. The operational principles of PIN diode-based designs were explained, including how a control current modulates the diode's RF resistance to vary attenuation, with more sophisticated bridge configurations enhancing performance. The chapter also introduced Field-Effect Transistors (FETs) as voltage-controlled alternatives, functioning as variable RF resistors.

In summary, this chapter has equipped you with the principles to understand, specify, and select RF attenuators. Mastery of the key specifications—including frequency range, attenuation range and resolution, insertion loss, power handling, and linearity—is essential for integrating these components into practical RF systems to achieve desired signal levels, protect sensitive circuitry, and ensure overall system stability and performance.

References

1. Tiwana M (2021) RF concepts, components and circuits for beginners. Udemy Inc., San Francisco, CA
2. Pozar DM (2011) Microwave engineering, 4th edn. John Wiley & Sons, Hoboken, NJ

The Software Defined Radios (SDRs) 15

Contents

15.1 Introduction to the SDR

Now, in order to understand the concept of software defined radio (SDR), consider the example of data communication link between the ground station and the UAV as shown in Fig. 15.1.

As you can see in Fig. 15.1, for that data communication we would need a radio communication device in the ground as well as the radio communication device in the UAV and we can use the software defined radio as this communication device and the next question is what is the software defined radio? A software defined radio is such a radio communication device in which the components that have been traditionally implemented in the hardware using different chips they are now implemented in the software on a general-purpose hardware [1].

M. Pakdel, *Understanding RF Systems*, Synthesis Lectures on RF/Microwaves,
https://doi.org/10.1007/978-3-032-19227-1_15

Fig. 15.1 The data communication link between ground station and UAV

15.2 The Advantages of SDR

Now, this approach of implementing the components of a wireless communication system in the software has many advantages, the first advantage is the flexibility and that means using the same generic hardware we can implement different type of communication systems by just changing the software. The second advantage is the rapid prototyping and that means using the same generic hardware we can implement new wireless communication systems very quickly by just changing the software. The third advantage is that the hardware does not become obsolete because the new wireless communication systems can still be implemented on that hardware by just changing the software. The fourth advantage is that the different wireless communication systems have different requirements in terms of range, transmit power and bit error, so we can design different wireless communication systems according to their requirements on the same hardware by adapting the software and as a result of all those facts the overall cost of the deployment of the wireless communication systems decreases substantially.

15.3 Generating PSK Using QAM as SDR Key Concept

Now, we will discuss about another key concept that how we can generate the PSK signal using the QAM. The equation of QAM is as below.

$$A\cos\left(\omega_0 t\right) - B\sin\left(\omega_0 t\right) = \sqrt{A^2 + B^2}\cos\left(\omega_0 t + \tan^{-1}\left(\frac{B}{A}\right)\right) \tag{15.1}$$

In Eq. (15.1), the I = A and Q = B for QAM signal, also the amplitude is $\sqrt{A^2 + B^2}$ and the phase Θ is equal to $\tan^{-1}\left(\frac{B}{A}\right)$ as depicted in Fig. 15.2.

We can see in Eq. (15.1) that we can vary the I and the Q in order to change the amplitude and also in order to change the phase Θ, now if we vary the I and Q in such a way that the amplitude ($\sqrt{I^2 + Q^2}$) remains constant but only the phase Θ changes between different symbols then we will end up in a situation like as illustrated in Fig. 15.3, and in this situation we can see that all the constellation points or all the symbols lie on the circle because their amplitudes are constant, in other words their energies are constant but only

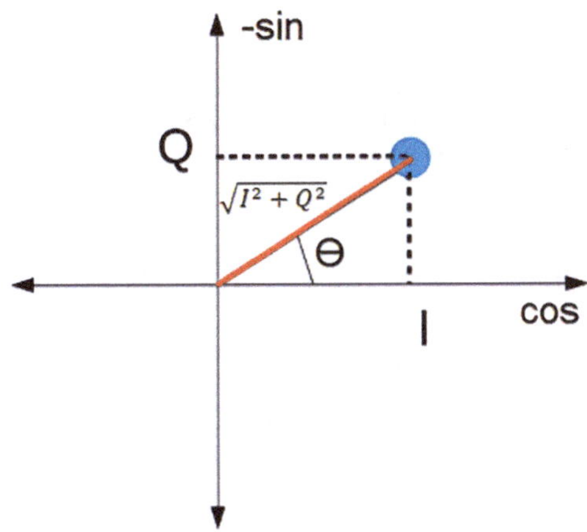

Fig. 15.2 The QAM phasor diagram

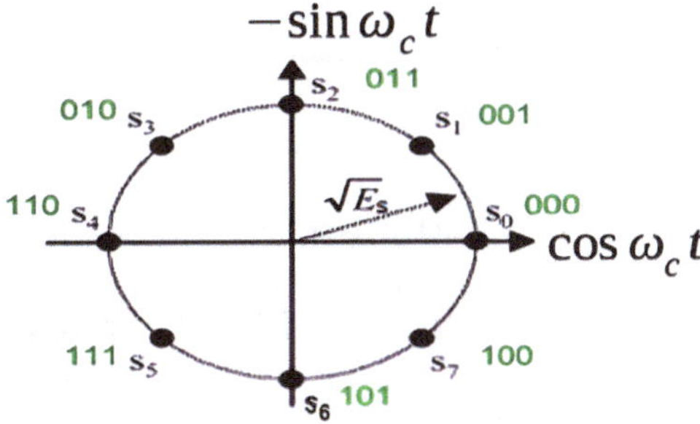

Fig. 15.3 The constellation diagram for the MPSK (M = 8) signal

their phases are changing, so that signal now in that form it is the MPSK signal or in other words we can generate the MPSK signal using the MQAM signal if we adjust the I and Q in such a way that the amplitude remains constant and only the phase changes so we can say that the phase shift keying is the special case of the quadrature amplitude modulation.

15.4 Generating ASK Using QAM as SDR Key Concept

We can also generate the ASK modulation using the QAM and we can do that by setting the Q component of the QAM as zero and by varying the I component to make different constellation points and we get the ASK modulation as shown in Fig. 15.4.

So, the ASK can be generated using the QAM by setting the Q component as zero. Similarly, we can see from the constellation diagrams in Fig. 15.5 that by varying I and Q in the QAM we can make different types of constellations.

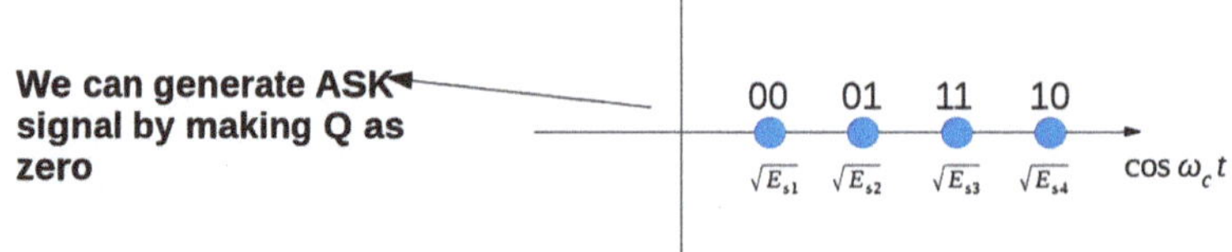

Fig. 15.4 The constellation diagram for the MASK (M = 4) signal

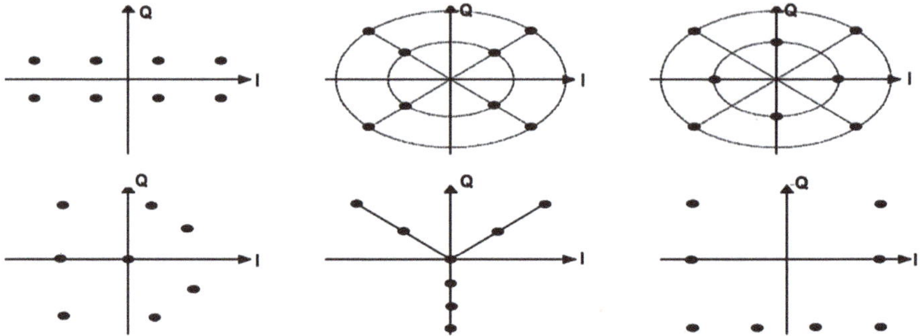

Fig. 15.5 The different types of constellations by varying I and Q in the QAM

15.5 Significance of QAM for SDRs

Now, the QAM is very important in the context of the SDRs, because by varying the I and the Q we can generate different constellation points that have different phase and that have different amplitude as depicted in Fig. 15.6.

So, we can use the QAM and vary the I and Q to generate different digital modulation schemes and the SDRs use the QAM to generate almost every type of digital as well as the analog modulation schemes now you may ask the question how the analog modulation is generated by the QAM which is a digital modulation scheme and we would come to this question later on.

15.6 Representing Symbols by Complex Numbers in SDR

Now, the complex numbers are important in the context of the SDR because the complex numbers are used to represent the symbols or in other words, they are used to represent the constellation points. For example, if we consider the symbol or constellation point it has the I and Q components, so in order to represent that constellation point or symbol, as a complex number we would write as S = I + Qj so, I is the real part and Q is the imaginary part as illustrated in Fig. 15.7.

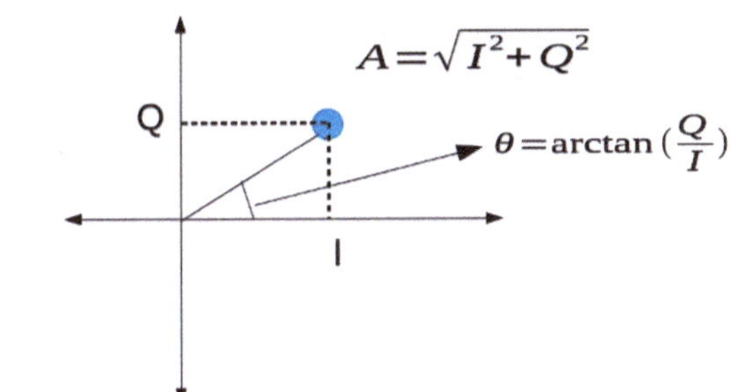

Fig. 15.6 The QAM phasor diagram

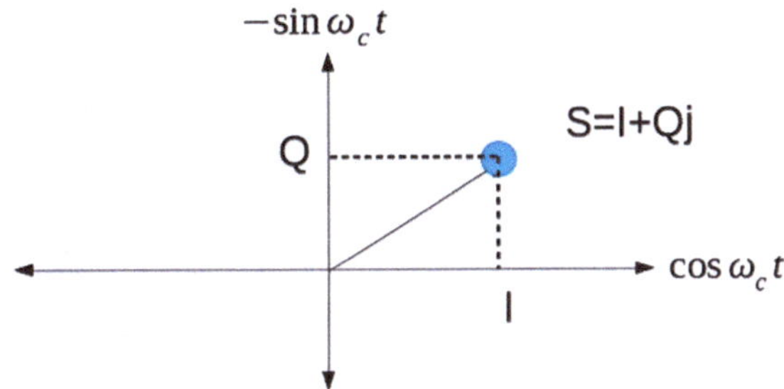

Fig. 15.7 The complex number to represent symbol in SDR

Now, if we have those 4 constellation points in Fig. 15.8, and we want to represent them mathematically using the complex numbers, in the first symbol S_0, 2 is the real part or it is the I and the Q is 1 so we would write it as $S_0 = 2 + j$.

Also, in the case of S_1, we can say −2 is the I and 1 is the Q, so we would write it as $S_1 = -2 + j$, and similarly S_2 can be written as $S_2 = -2\text{-}j$ and in the case of S_3, 2 is the I and − 1 is the Q so we write it as $S_3 = 2\text{-}j$, so that is how the symbols are represented using the complex numbers and that is why in this representation these symbols are called as the complex symbols.

15.7 The SDR Transmitter Block Diagram

Now, we come to the block diagram of SDR and in this section we are discussing about the transmitter side of SDR, now here we want to point out that the general transmitter block diagram of SDR and almost all the SDRs follow this diagram however there may be slight differences in how that diagram is implemented in different SDRs as shown in Fig. 15.9.

As you can see in Fig. 15.9, we have the personal computer (PC) that is running the GNU radio software and that PC is connected to the SDR which is called as the RF front end, now all the baseband processing is done in the GNU radio, while all the baseband processing or the baseband modulation is done in the SDR and that is why SDR is also called as the RF front end. Now, suppose the symbol $S_0 = 2 + 3j$ needs to be modulated and transmitted using the SDR, so the GNU radio would generate that complex symbol that

Fig. 15.8 The example of 4 constellation points

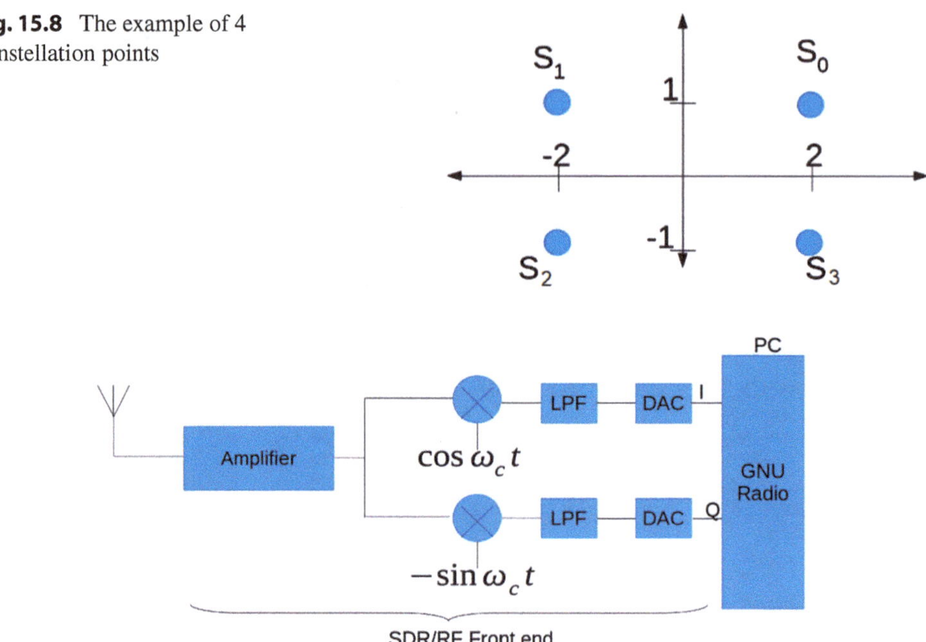

Fig. 15.9 The general block diagram of SDR transmitter

has the I part and also that has a Q part, The I part is given to the upper branch in Fig. 15.9 and the Q part would be given to the lower branch so that I is in digital form and using the digital to analog converter (DAC), it would be converted to the analog form and it would pass through the lowpass filter (LPF) and then it would be modulated using the cosine carrier, similarly that 3 would be converted from digital to analog form using the DAC and it would pass through the LPF and then it would be modulated using the sine carrier and then we would add those two signals, so now that complex is in its modulated form and that modulated complex symbol would then be given to the amplifier that would amplify the modulated complex symbol and then it would send it to the wireless channel using that antenna [2].

15.8 The SDR Receiver Block Diagram

Now, that complex symbol in the previous section would be received by the receiving SDR in Fig. 15.10 and then it would pass through the amplifier and after that the received symbol is demodulated and for the demodulation, we have two demodulators one is the I demodulator and other is the Q demodulator.

In the I demodulator the received symbol will be multiplied by the cosine carrier and then it would pass through the LPF and at the output of the LPF and for simplicity we would assume that there is no noise in the channel we would get 2 and that is in analog

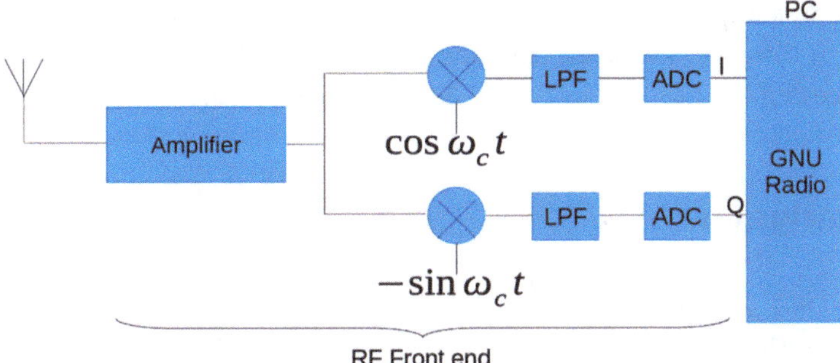

Fig. 15.10 The general block diagram of SDR receiver

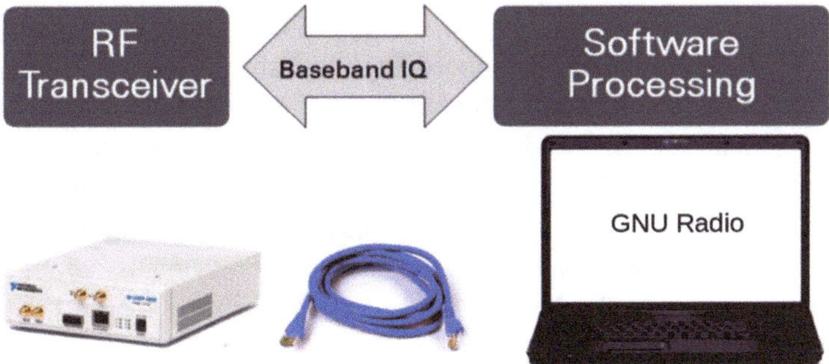

Fig. 15.11 The simplified setup of an SDR

form and it would be converted to the digital form by the analog to digital converter (ADC) and then it would be given to the GNU radio software, now in the Q branch the received complex symbol would be multiplied by the carrier -sin(ω_ct), and then it would pass through the LPF and there we would get 3 and that would then be digitized using the ADC and then it would be given to the GNU radio, now the GNU radio has received the complex symbol $S_0 = 2 + 3j$ and that symbol is now in the baseband and after that the GNU radio would decode that symbol and convert that symbol to the corresponding bits [2].

15.9 The Simplified Setup of an SDR

Now, the simplified setup of an SDR is depicted in Fig. 15.11.

As you can see in Fig. 15.11 we have a PC that is running the GNU radio and all the baseband processing is done in the GNU radio and that PC is connected to the SDR and that connection can be through an Ethernet or it can be through the USB, now that SDR is serving as the RF transmitter and RF receiver or in other words that SDR is serving as the RF transceiver and it receives the I and Q in the baseband from the GNU radio as complex symbols and it modulates those symbols and send them to the wireless channel and from the wireless channel it receives the complex symbols, demodulates those complex symbols and gives those demodulated complex symbols as I and Q to the GNU radio where they are processed by the GNU radio and decoded into the bits [1].

15.10 Conclusion

This chapter has provided a comprehensive introduction to Software Defined Radio (SDR), a transformative technology that redefines the architecture of wireless communication systems. We began by framing SDR's core principle: moving key signal processing functions from fixed, application-specific hardware into reconfigurable software running on a generic hardware platform, such as a PC connected to an RF front-end.

The chapter detailed the compelling advantages of this paradigm, including unprecedented flexibility, rapid prototyping, reduced hardware obsolescence, and significant cost reduction in system deployment. A critical technical foundation was established by demonstrating that Quadrature Amplitude Modulation (QAM) serves as a universal engine for digital modulation within SDRs. We showed how Phase Shift Keying (PSK) and Amplitude Shift Keying (ASK) are generated as specific cases of QAM by manipulating the in-phase (I) and quadrature (Q) components, and how constellation symbols are naturally represented as complex numbers ($S = I + jQ$).

Finally, the theoretical concepts were connected to practical implementation through detailed transmitter and receiver block diagrams. We traced the signal path from complex symbol generation in software (e.g., GNU Radio), through digital-to-analog conversion, modulation, and transmission, and back through reception, demodulation, analog-to-digital conversion, and software-based decoding. This illustrated the clear separation between the flexible, software-defined baseband processing and the essential, generic RF hardware.

In summary, this chapter has equipped you with the principles to understand how SDRs decouple radio functionality from hardware, enabling a single device to emulate countless radios through software alone. Mastery of the QAM foundation, complex symbol representation, and system architecture is essential for engaging with modern and future wireless technologies, where adaptability and software control are paramount.

References

1. Tiwana M (2021) RF concepts, components and circuits for beginners. Udemy Inc., San Francisco, CA
2. Collins T, Getz R, Pu D, Wyglinski AM (2018) Software defined Radio for engineers. Artech House

The RF Design with DeepSeek 16

Contents

16.1 Introduction to the DeepSeek

DeepSeek is a state-of-the-art artificial intelligence (AI) model designed to assist in a wide range of computational tasks, including coding, data analysis, and problem-solving [1]. Its advanced natural language processing capabilities make it an invaluable tool for engineers and researchers working on complex projects, such as radio frequency (RF) design. By leveraging DeepSeek, users can generate, debug, and optimize Python code efficiently, enabling rapid prototyping and simulation of RF circuits and systems. In this chapter, we explore how DeepSeek can be integrated into the RF design workflow using Python and Google Colab. Google Colab, a cloud-based Jupyter notebook environment, provides a seamless platform for running Python code without the need for local software installation [2]. Combined with DeepSeek's ability to generate and refine code, this setup allows engineers to focus on high-level design concepts while automating repetitive coding tasks. Whether you are designing filters, amplifiers, or antennas, DeepSeek can help streamline the process, from initial concept to simulation and analysis. The following sections will demonstrate how to use DeepSeek to write Python scripts for RF design tasks and how to execute these scripts in Google Colab. By the end of this chapter, you will have a practical understanding of how to harness the power of DeepSeek and Google Colab to enhance your RF design workflow [1, 2].

M. Pakdel, *Understanding RF Systems*, Synthesis Lectures on RF/Microwaves,
https://doi.org/10.1007/978-3-032-19227-1_16

16.2 Designing the QAM with Python (Colab)

Designing a Quadrature Amplitude Modulation (QAM) system in Python using Google Colab involves several steps. Below is a step-by-step guide to implement a basic QAM modulator and demodulator. First, we import the necessary libraries.

```
import numpy as np
import matplotlib.pyplot as plt
from scipy import signal
```

Step-by-Step Breakdown is as follows:

1. Initialization and Parameters

```
num_symbols = 1000           # How many symbols to send
bits_per_symbol = 4          # 16-QAM = 4 bits per symbol
samples_per_symbol = 10       # How many samples represent
1 symbol
fs = 1000                    # Sampling frequency (Hz)
f_c = 100                    # Carrier frequency (Hz)
T = 1/fs                     # Time between samples
```

Think of this as setting up the "rules" of communication, ach symbol carries 4 bits of information, and we oversample (10 samples per symbol) to make filtering easier.

2. Data Generation and Symbol Mapping

```
# Generate random bits
data     =     np.random.randint(0,     2,     num_symbols     *
bits_per_symbol)  # 4000 random bits

# Reshape into symbols
# 4000 bits → 1000 symbols × 4 bits each
symbols = data.reshape((num_symbols, bits_per_symbol))
```

For example:

```
Original bits: [1, 0, 1, 1, 0, 1, 0, 0, ...]
Reshaped: [[1, 0, 1, 1], → Symbol 1 (bits: 1011)
      [0, 1, 0, 0], → Symbol 2 (bits: 0100)
      ...]
```

3. Create 16-QAM Constellation

```
constellation = np.array([
    -3-3j, -3-1j, -3+3j, -3+1j,   # Row 1
    -1-3j, -1-1j, -1+3j, -1+1j,   # Row 2
    3-3j,  3-1j,  3+3j,  3+1j,    # Row 3
    1-3j,  1-1j,  1+3j,  1+1j     # Row 4
]) / np.sqrt(10)  # Normalize
```

Visual representation (I vs Q plane):

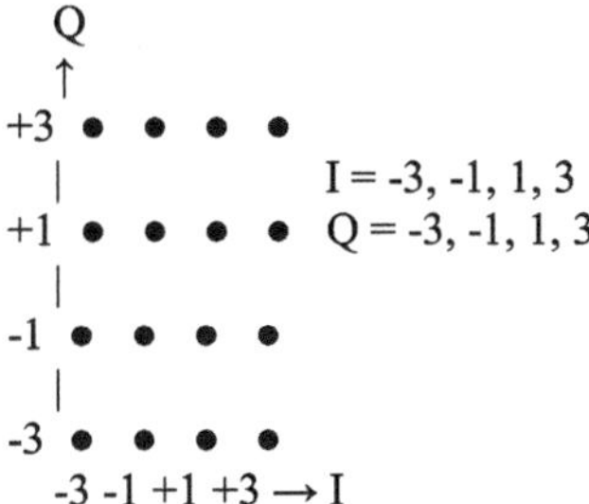

Each dot = a possible symbol (16 total).
Each symbol = 4 bits mapped to (I, Q) coordinates.
Normalization ensures constant average power.

4. Map Bits to Constellation Points

```
# Convert binary to decimal (0-15)
symbols_decimal = [sum(symbol[i] * 2**(3-i) for i in range(4))
                   for symbol in symbols]

# Map to constellation
symbols_complex = np.array([constellation[val] for val in symbols_decimal])
```

For example:

Bits: [1, 0, 1, 1] → Decimal: 1×8 + 0×4 + 1×2 + 1×1 = 11
Constellation[11] = 3+1j (normalized)
So: I = 3/√10, Q = 1/√10

5. Upsampling and Modulation

```
# Upsample: 1 sample per symbol → 10 samples per symbol
I_up = np.repeat(I_symbols, samples_per_symbol)  # Stretch I
Q_up = np.repeat(Q_symbols, samples_per_symbol)  # Stretch Q

# Create carriers
carrier_I = np.cos(2πf_c t)   # Cosine wave at 100Hz
carrier_Q = -np.sin(2πf_c t)  # Negative sine wave at 100Hz

# Modulate
I_mod = I_up * carrier_I   # I component modulates cosine
Q_mod = Q_up * carrier_Q   # Q component modulates sine

# Combine
modulated_signal = I_mod + Q_mod
```

6. Demodulation Process

```
# Multiply received signal by carriers
I_demod = modulated_signal * 2 * carrier_I
Q_demod = modulated_signal * 2 * carrier_Q
```

Why multiply by 2? The math: $\cos^2(x) = ½ + ½\cos(2x)$, after low-pass filtering: ½ remains, so we multiply by 2 to get original amplitude.

7. Low-Pass Filtering

```
cutoff_freq = f_c / (fs/2)  # Normalized cutoff (100Hz /
500Hz = 0.2)
b, a = signal.butter(6, cutoff_freq)
I_filtered = signal.lfilter(b, a, I_demod)
Q_filtered = signal.lfilter(b, a, Q_demod)
```

Before filtering:

Mixed signal = Baseband + 2×Carrier frequency

After filtering:

Remove high-frequency components (2×f_c), and keep only baseband (original I and Q)

8. Downsampling and Symbol Recovery

```
# Take every 10th sample (back to 1 sample per symbol)
I_recovered = I_filtered[::samples_per_symbol]
Q_recovered = Q_filtered[::samples_per_symbol]

# Remove filter delay
I_recovered = I_recovered[delay:]
Q_recovered = Q_recovered[delay:]
```

9. Decision Making (Nearest Neighbor)

```
for recv_symbol in recovered_complex:
    # Calculate distance to all 16 constellation points
    distances = np.abs(constellation - recv_symbol)
    # Find closest one
    closest_idx = np.argmin(distances)
    decoded_symbols.append(closest_idx)
```

For example:

Received point: (2.8, 0.9)

Calculate distances to all 16 points:

Distance to (-3,-3) = $\sqrt{((2.8+3)^2 + (0.9+3)^2)} \approx 6.1$
Distance to (3,1) = $\sqrt{((2.8-3)^2 + (0.9-1)^2)} \approx 0.22$ ← Closest!
Distance to (1,3) = $\sqrt{((2.8-1)^2 + (0.9-3)^2)} \approx 2.4$
...
Choose (3,1) → Index 11 → Bits: 1011

10. Convert Back to Bits

```
for symbol in decoded_symbols:
    # Convert decimal 0-15 to 4 bits
    bits = [(symbol >> (3-i)) & 1 for i in range(4)]
    recovered_bits.extend(bits)
```

For example:

Symbol index = 11 (decimal)
Binary: 11 = 1011 (1×8 + 0×4 + 1×2 + 1×1)
Extract bits: [(11 >> 3) & 1 = 1,
(11 >> 2) & 1 = 0,
(11 >> 1) & 1 = 1,
(11 >> 0) & 1 = 1]

11. Error Calculation

```
bit_errors = np.sum(original_bits != recovered_bits)
ber = bit_errors / total_bits
```

The code compares original vs. recovered bits, and calculates Bit Error Rate (BER). For perfect recovery we have 0 errors.

Here is the full code:

```
import numpy as np
import matplotlib.pyplot as plt
from scipy import signal

# Parameters
num_symbols = 1000  # Number of symbols to transmit
bits_per_symbol = 4  # 16-QAM (4 bits per symbol)
samples_per_symbol = 10  # Samples per symbol
fs = 1000  # Sampling frequency (Hz)
T = 1 / fs  # Sampling period
f_c = 100  # Carrier frequency (Hz)

# Generate random binary data
data = np.random.randint(0, 2, num_symbols *
bits_per_symbol)

# Reshape data into symbols (each symbol has 4 bits)
symbols = data.reshape((num_symbols, bits_per_symbol))

# Convert binary to decimal for each symbol (0-15)
symbols_decimal = np.zeros(num_symbols, dtype=int)
for i in range(num_symbols):
    for j in range(bits_per_symbol):
        symbols_decimal[i] += symbols[i, j] * (2 **
(bits_per_symbol - 1 - j))

# 16-QAM constellation mapping (Gray coded)
# Create constellation points for 16-QAM
constellation = np.array([
    -3-3j, -3-1j, -3+3j, -3+1j,
    -1-3j, -1-1j, -1+3j, -1+1j,
    3-3j,  3-1j,  3+3j,  3+1j,
    1-3j,  1-1j,  1+3j,  1+1j
]) / np.sqrt(10)  # Normalize for unit average power

# Map decimal symbols to constellation points
symbols_complex = np.array([constellation[val] for val in
symbols_decimal])

# Separate I and Q components
I_symbols = np.real(symbols_complex)
Q_symbols = np.imag(symbols_complex)

# Time vector
t_symbol = np.arange(num_symbols * samples_per_symbol) *
T

# Upsample symbols to match the sampling rate
I_up = np.repeat(I_symbols, samples_per_symbol)
Q_up = np.repeat(Q_symbols, samples_per_symbol)
```

```
# Carrier signals
carrier_I = np.cos(2 * np.pi * f_c * t_symbol)
carrier_Q = -np.sin(2 * np.pi * f_c * t_symbol)  #
Negative for Q

# Modulate the signal
I_mod = I_up * carrier_I
Q_mod = Q_up * carrier_Q

# Combine I and Q components
modulated_signal = I_mod + Q_mod

# Plot the modulated signal
plt.figure(figsize=(12, 8))

plt.subplot(2, 2, 1)
plt.plot(t_symbol[:200], modulated_signal[:200])
plt.title('Modulated Signal (16-QAM) - First 200
samples')
plt.xlabel('Time (s)')
plt.ylabel('Amplitude')
plt.grid(True)

# Plot constellation
plt.subplot(2, 2, 2)
plt.scatter(I_symbols, Q_symbols, alpha=0.6)
plt.title('16-QAM Constellation (Transmitted)')
plt.xlabel('In-phase (I)')
plt.ylabel('Quadrature (Q)')
plt.grid(True)
plt.axis('equal')

# Demodulate the signal
I_demod = modulated_signal * 2 * carrier_I  # Multiply by
2 to compensate for 0.5 factor
Q_demod = modulated_signal * 2 * carrier_Q  # Multiply by
2 to compensate for 0.5 factor

# Low-pass filter design
cutoff_freq = f_c / (fs/2)  # Cutoff frequency normalized
by Nyquist
b, a = signal.butter(6, cutoff_freq)

# Apply filters
I_filtered = signal.lfilter(b, a, I_demod)
Q_filtered = signal.lfilter(b, a, Q_demod)

# Downsample to get the original symbols
I_recovered = I_filtered[::samples_per_symbol]
Q_recovered = Q_filtered[::samples_per_symbol]
```

```
# Remove edge effects (filter delay)
delay = 3  # Estimated filter delay
I_recovered = I_recovered[delay:]
Q_recovered = Q_recovered[delay:]
symbols_complex = symbols_complex[:-delay]  # Adjust
original symbols for comparison

# Combine I and Q into complex symbols
recovered_complex = I_recovered + 1j * Q_recovered

# Plot recovered constellation
plt.subplot(2, 2, 3)
plt.scatter(I_recovered, Q_recovered, alpha=0.6,
color='red')
plt.title('16-QAM Constellation (Received)')
plt.xlabel('In-phase (I)')
plt.ylabel('Quadrature (Q)')
plt.grid(True)
plt.axis('equal')

# Plot eye diagram
plt.subplot(2, 2, 4)
offset = samples_per_symbol // 2
for i in range(20):  # Plot first 20 symbols
    start = i * samples_per_symbol + offset
    end = start + samples_per_symbol

plt.plot(np.arange(samples_per_symbol)/samples_per_symbol
,
             I_filtered[start:end], 'b-', alpha=0.5)
plt.title('Eye Diagram (I component)')
plt.xlabel('Symbol period')
plt.ylabel('Amplitude')
plt.grid(True)

plt.tight_layout()
plt.show()

# Decision: Find closest constellation point for each
received symbol
decoded_symbols = []
for recv_symbol in recovered_complex:
    # Calculate distances to all constellation points
    distances = np.abs(constellation - recv_symbol)
    # Find index of closest constellation point
    closest_idx = np.argmin(distances)
    decoded_symbols.append(closest_idx)

# Convert decoded symbols back to bits
recovered_bits = []
for symbol in decoded_symbols:
    # Convert decimal symbol (0-15) to 4 bits
```

```
    bits = [(symbol >> (bits_per_symbol - 1 - i)) & 1 for
i in range(bits_per_symbol)]
    recovered_bits.extend(bits)

# Convert to numpy array
recovered_bits = np.array(recovered_bits)

# Trim original data to match length (due to filter
delay)
original_bits_trimmed = data[:-delay *
bits_per_symbol].flatten()

# Verify the data
bit_errors = np.sum(original_bits_trimmed !=
recovered_bits[:len(original_bits_trimmed)])
total_bits = len(original_bits_trimmed)
ber = bit_errors / total_bits

print(f"Total bits transmitted: {total_bits}")
print(f"Bit errors: {bit_errors}")
print(f"Bit Error Rate (BER): {ber:.6f}")

if bit_errors == 0:
    print("Data successfully recovered!")
else:
    print(f"Data recovery has {bit_errors} errors out of
{total_bits} bits")
    print(f"Accuracy: {(1-ber)*100:.2f}%")

# Plot bit error comparison
plt.figure(figsize=(10, 4))
plt.plot(original_bits_trimmed[:100], 'bo-',
label='Original bits', markersize=8, alpha=0.7)
plt.plot(recovered_bits[:100], 'rx-', label='Recovered
bits', markersize=6, alpha=0.7)
plt.title('Bit Comparison (First 100 bits)')
plt.xlabel('Bit index')
plt.ylabel('Bit value')
plt.legend()
plt.grid(True)
plt.show()
```

```
Total bits transmitted: 3988
Bit errors: 1957
Bit Error Rate (BER): 0.490722
Data recovery has 1957 errors out of 3988 bits
Accuracy: 50.93%
```

```
import numpy as np
import matplotlib.pyplot as plt
from scipy import signal
```

The 16-QAM modulated signal, transmitted and received constellation, as well as the eye diagram (I component) are shown in Fig. 16.1.

Also, the bit comparison diagram for first 100 bits is depicted in Fig 16.2.

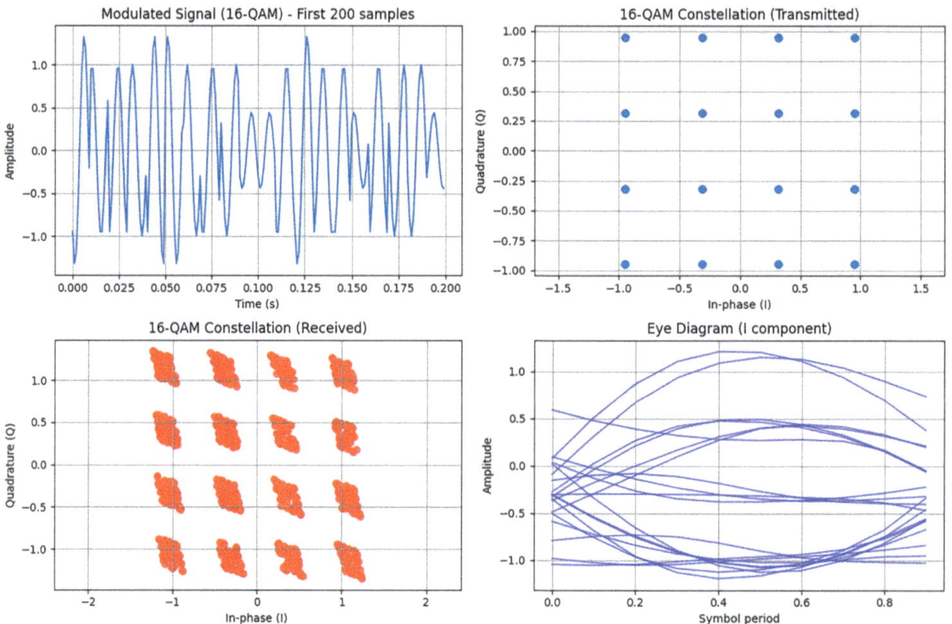

Fig. 16.1 The 16-QAM modulated signal, transmitted and received constellation, and eye diagram (I component)

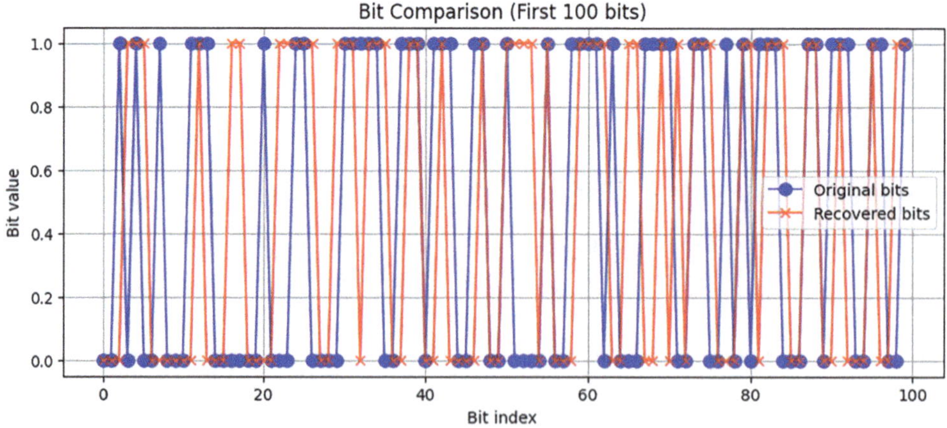

Fig. 16.2 The bit comparison diagram for first 100 bits

16.3 Designing RF Filters with Python (Colab)

Designing LC filters (low-pass, high-pass, bandpass, and bandstop) in Python using Google Colab involves calculating the component values (inductors and capacitors) based on the desired cutoff frequencies and plotting their frequency responses. Below is a step-by-step guide to designing these filters.

Step 1: Import Required Libraries

```
# Filter parameters
R = 50 # Impedance (Ohms)
f_low = 1e6 # Low cutoff frequency (1 MHz)
f_high = 10e6 # High cutoff frequency (10 MHz)
order = 2 # Filter order
```

Step 2: Define Filter Parameters

Define the filter parameters such as cutoff frequencies, order, and impedance.

```
# Low-pass filter design
def lc_lowpass(R, f_c, order):
    L = R / (2 * np.pi * f_c)
    C = 1 / (R * 2 * np.pi * f_c)
    return L, C
L_lp, C_lp = lc_lowpass(R, f_low, order)
print(f"Low-Pass  Filter:  L  =  {L_lp:.6f}  H,  C  =
{C_lp:.12f} F")
# Transfer function
num_lp = [1]
den_lp = [L_lp * C_lp, L_lp / R, 1]
system_lp = signal.TransferFunction(num_lp, den_lp)
w_lp,   mag_lp,   phase_lp   =   signal.bode(system_lp,
np.logspace(5, 8, 1000))
# Plot frequency response
plt.figure(figsize=(10, 6))
plt.semilogx(w_lp, mag_lp)
plt.title('Low-Pass Filter Frequency Response')
plt.xlabel('Frequency (rad/s)')
plt.ylabel('Magnitude (dB)')
plt.grid(True)
plt.show()
```

Step 3: Design Low-Pass Filter

A low-pass filter allows frequencies below a cutoff frequency to pass through.

```
# High-pass filter design
def lc_highpass(R, f_c, order):
    L = R / (2 * np.pi * f_c)
    C = 1 / (R * 2 * np.pi * f_c)
    return L, C
L_hp, C_hp = lc_highpass(R, f_high, order)
print(f"High-Pass  Filter:  L  =  {L_hp:.6f}  H,  C  =
{C_hp:.12f} F")
# Transfer function
num_hp = [L_hp * C_hp, 0, 0]
den_hp = [L_hp * C_hp, L_hp / R, 1]
system_hp = signal.TransferFunction(num_hp, den_hp)
w_hp,   mag_hp,   phase_hp   =   signal.bode(system_hp,
np.logspace(5, 8, 1000))
# Plot frequency response
plt.figure(figsize=(10, 6))
plt.semilogx(w_hp, mag_hp)
plt.title('High-Pass Filter Frequency Response')
plt.xlabel('Frequency (rad/s)')
plt.ylabel('Magnitude (dB)')
plt.grid(True)
plt.show()
```

Step 4: Design High-Pass Filter

A high-pass filter allows frequencies above a cutoff frequency to pass through.

```
# Bandpass filter design
def lc_bandpass(R, f_low, f_high, order):
    f_center = np.sqrt(f_low * f_high)
    bandwidth = f_high - f_low
    L1 = R / (2 * np.pi * bandwidth)
    C1 = 1 / (R * 2 * np.pi * bandwidth)
    L2 = R / (2 * np.pi * f_center)
    C2 = 1 / (R * 2 * np.pi * f_center)
    return L1, C1, L2, C2
L1_bp, C1_bp, L2_bp, C2_bp = lc_bandpass(R, f_low, f_high,
order)
print(f"Bandpass  Filter:  L1  =  {L1_bp:.6f}  H,  C1  =
{C1_bp:.12f} F, L2 = {L2_bp:.6f} H, C2 = {C2_bp:.12f} F")
# Transfer function
num_bp = [L1_bp * C1_bp, 0, 0]
den_bp = [L1_bp * C1_bp * L2_bp * C2_bp, L1_bp * C1_bp *
L2_bp / R, L1_bp * C1_bp + L2_bp * C2_bp, L2_bp / R, 1]
system_bp = signal.TransferFunction(num_bp, den_bp)
w_bp,   mag_bp,   phase_bp   =   signal.bode(system_bp,
np.logspace(5, 8, 1000))
# Plot frequency response
plt.figure(figsize=(10, 6))
plt.semilogx(w_bp, mag_bp)
plt.title('Bandpass Filter Frequency Response')
plt.xlabel('Frequency (rad/s)')
plt.ylabel('Magnitude (dB)')
plt.grid(True)
plt.show()
```

Step 5: Design Bandpass Filter

A bandpass filter allows frequencies within a specific range to pass through.

```
# Bandstop filter design
def lc_bandstop(R, f_low, f_high, order):
    f_center = np.sqrt(f_low * f_high)
    bandwidth = f_high - f_low
    L1 = R / (2 * np.pi * bandwidth)
    C1 = 1 / (R * 2 * np.pi * bandwidth)
    L2 = R / (2 * np.pi * f_center)
    C2 = 1 / (R * 2 * np.pi * f_center)
    return L1, C1, L2, C2
L1_bs, C1_bs, L2_bs, C2_bs = lc_bandstop(R, f_low,
f_high, order)
print(f"Bandstop Filter: L1 = {L1_bs:.6f} H, C1 =
{C1_bs:.12f} F, L2 = {L2_bs:.6f} H, C2 = {C2_bs:.12f}F")
# Transfer function
num_bs = [L1_bs * C1_bs * L2_bs * C2_bs, 0, L1_bs * C1_bs
+ L2_bs * C2_bs, 0, 1]
den_bs = [L1_bs * C1_bs * L2_bs * C2_bs, L1_bs * C1_bs *
L2_bs / R, L1_bs * C1_bs + L2_bs * C2_bs, L2_bs / R, 1]
system_bs = signal.TransferFunction(num_bs, den_bs)
w_bs, mag_bs, phase_bs = signal.bode(system_bs,
np.logspace(5, 8, 1000))
# Plot frequency response
plt.figure(figsize=(10, 6))
plt.semilogx(w_bs, mag_bs)
plt.title('Bandstop Filter Frequency Response')
plt.xlabel('Frequency (rad/s)')
plt.ylabel('Magnitude (dB)')
plt.grid(True)
plt.show()
```

Step 6: Design Bandstop Filter

A bandstop filter blocks frequencies within a specific range.

```
import numpy as np
import matplotlib.pyplot as plt
from scipy import signal

# Filter parameters
R = 50 # Impedance (Ohms)
f_low = 1e6 # Low cutoff frequency (1 MHz)
f_high = 10e6 # High cutoff frequency (10 MHz)
order = 2 # Filter order

# Low-pass filter design
def lc_lowpass(R, f_c, order):
    L = R / (2 * np.pi * f_c)
    C = 1 / (R * 2 * np.pi * f_c)
    return L, C

L_lp, C_lp = lc_lowpass(R, f_low, order)
print(f"Low-Pass Filter: L = {L_lp:.6f} H, C =
{C_lp:.12f} F")

# Transfer function
num_lp = [1]
den_lp = [L_lp * C_lp, L_lp / R, 1]
system_lp = signal.TransferFunction(num_lp, den_lp)
w_lp, mag_lp, phase_lp = signal.bode(system_lp,
np.logspace(5, 8, 1000))

# Plot frequency response
plt.figure(figsize=(10, 6))
plt.semilogx(w_lp, mag_lp)
plt.title('Low-Pass Filter Frequency Response')
plt.xlabel('Frequency (rad/s)')
plt.ylabel('Magnitude (dB)')
plt.grid(True)
plt.show()
```

Explanation of the code:

Low-Pass Filter: Allows frequencies below the cutoff frequency to pass.

High-Pass Filter: Allows frequencies above the cutoff frequency to pass.

Band-pass Filter: Allows frequencies within a specific range to pass.

Band-stop Filter: Blocks frequencies within a specific range.

Each filter is designed using LC components, and the frequency response is plotted using the scipy.signal library. You can adjust the parameters (e.g., cutoff frequencies, impedance) to suit your requirements.

Full Code: Here is the full code for the lowpass, high pass, band pass and band stop filters.

```
Low-Pass Filter: L = 0.000008 H, C = 0.000000003183 F
```

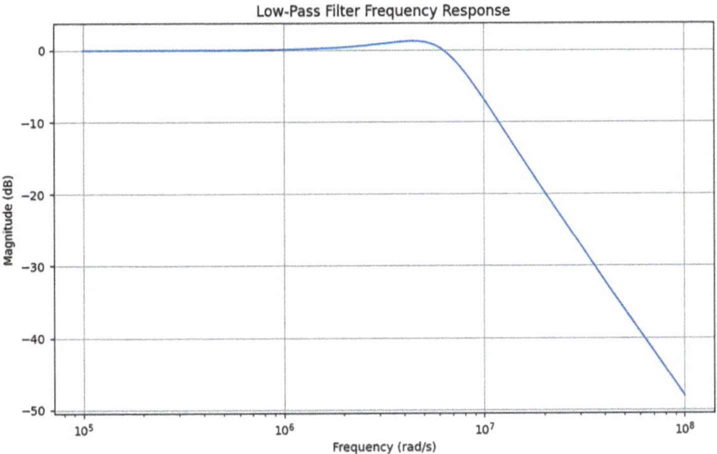

Fig. 16.3 The resulted low-pass filter frequency response

The resulted low-pass filter frequency response is shown in Fig. 16.3.

```
# High-pass filter design
def lc_highpass(R, f_c, order):
    L = R / (2 * np.pi * f_c)
    C = 1 / (R * 2 * np.pi * f_c)
    return L, C

L_hp, C_hp = lc_highpass(R, f_high, order)
print(f"High-Pass Filter: L = {L_hp:.6f} H, C =
{C_hp:.12f} F")

# Transfer function
num_hp = [L_hp * C_hp, 0, 0]
den_hp = [L_hp * C_hp, L_hp / R, 1]
system_hp = signal.TransferFunction(num_hp, den_hp)
w_hp, mag_hp, phase_hp = signal.bode(system_hp,
np.logspace(5, 8, 1000))

# Plot frequency response
plt.figure(figsize=(10, 6))
plt.semilogx(w_hp, mag_hp)
plt.title('High-Pass Filter Frequency Response')
plt.xlabel('Frequency (rad/s)')
plt.ylabel('Magnitude (dB)')
plt.grid(True)
plt.show()
```

The resulted high-pass filter frequency response is illustrated in Fig. 16.4.

```
High-Pass Filter: L = 0.000001 H, C = 0.000000000318 F
```

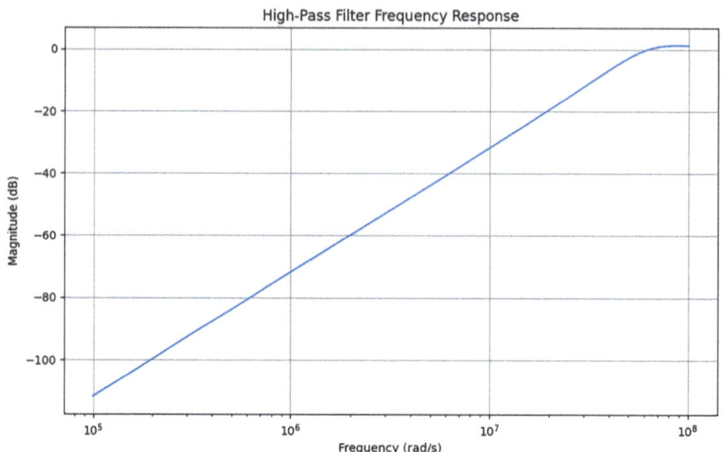

Fig. 16.4 The resulted high-pass filter frequency response

```
# Bandpass filter design
def lc_bandpass(R, f_low, f_high, order):
    f_center = np.sqrt(f_low * f_high)
    bandwidth = f_high - f_low
    L1 = R / (2 * np.pi * bandwidth)
    C1 = 1 / (R * 2 * np.pi * bandwidth)
    L2 = R / (2 * np.pi * f_center)
    C2 = 1 / (R * 2 * np.pi * f_center)
    return L1, C1, L2, C2

L1_bp, C1_bp, L2_bp, C2_bp = lc_bandpass(R, f_low,
f_high, order)
print(f"Bandpass Filter: L1 = {L1_bp:.6f} H, C1 =
{C1_bp:.12f} F, L2 = {L2_bp:.6f} H, C2 = {C2_bp:.12f} F")

# Transfer function
num_bp = [L1_bp * C1_bp, 0, 0]
den_bp = [L1_bp * C1_bp * L2_bp * C2_bp, L1_bp * C1_bp *
L2_bp / R, L1_bp * C1_bp + L2_bp * C2_bp, L2_bp / R, 1]
system_bp = signal.TransferFunction(num_bp, den_bp)
w_bp, mag_bp, phase_bp = signal.bode(system_bp,
np.logspace(5, 8, 1000))

# Plot frequency response
plt.figure(figsize=(10, 6))
plt.semilogx(w_bp, mag_bp)
plt.title('Bandpass Filter Frequency Response')
plt.xlabel('Frequency (rad/s)')
plt.ylabel('Magnitude (dB)')
plt.grid(True)
plt.show()
```

The resulted bandpass filter frequency response is shown in Fig. 16.5.

```
Bandpass Filter: L1 = 0.000001 H, C1 = 0.000000000354 F,
L2 = 0.000003 H, C2 = 0.000000001007 F
```

```
# Bandstop filter design
def lc_bandstop(R, f_low, f_high, order):
    f_center = np.sqrt(f_low * f_high)
    bandwidth = f_high - f_low
    L1 = R / (2 * np.pi * bandwidth)
    C1 = 1 / (R * 2 * np.pi * bandwidth)
    L2 = R / (2 * np.pi * f_center)
    C2 = 1 / (R * 2 * np.pi * f_center)
    return L1, C1, L2, C2

L1_bs, C1_bs, L2_bs, C2_bs = lc_bandstop(R, f_low,
f_high, order)
print(f"Bandstop Filter: L1 = {L1_bs:.6f} H, C1 =
{C1_bs:.12f} F, L2 = {L2_bs:.6f} H, C2 = {C2_bs:.12f} F")

# Transfer function
num_bs = [L1_bs * C1_bs * L2_bs * C2_bs, 0, L1_bs * C1_bs
+ L2_bs * C2_bs, 0, 1]
den_bs = [L1_bs * C1_bs * L2_bs * C2_bs, L1_bs * C1_bs *
L2_bs / R, L1_bs * C1_bs + L2_bs * C2_bs, L2_bs / R, 1]
system_bs = signal.TransferFunction(num_bs, den_bs)
w_bs, mag_bs, phase_bs = signal.bode(system_bs,
np.logspace(5, 8, 1000))

# Plot frequency response
plt.figure(figsize=(10, 6))
plt.semilogx(w_bs, mag_bs)
plt.title('Bandstop Filter Frequency Response')
plt.xlabel('Frequency (rad/s)')
plt.ylabel('Magnitude (dB)')
plt.grid(True)
plt.show()
```

The resulted bandstop filter frequency response is depicted in Fig. 16.6.

```
Bandstop Filter: L1 = 0.000001 H, C1 = 0.000000000354 F,
L2 = 0.000003 H, C2 = 0.000000001007 F
```

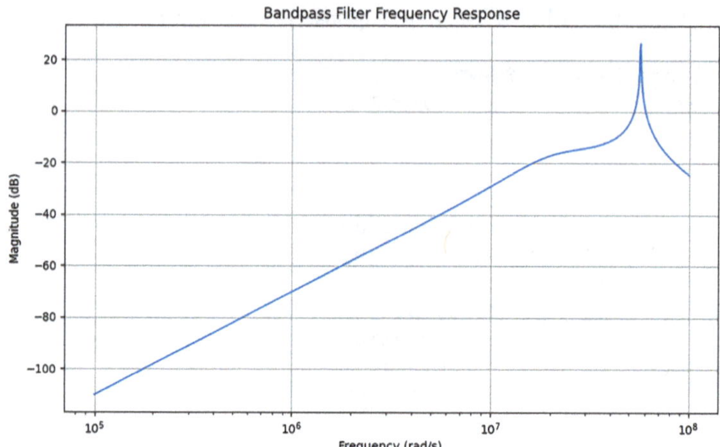

Fig. 16.5 The resulted bandpass filter frequency response

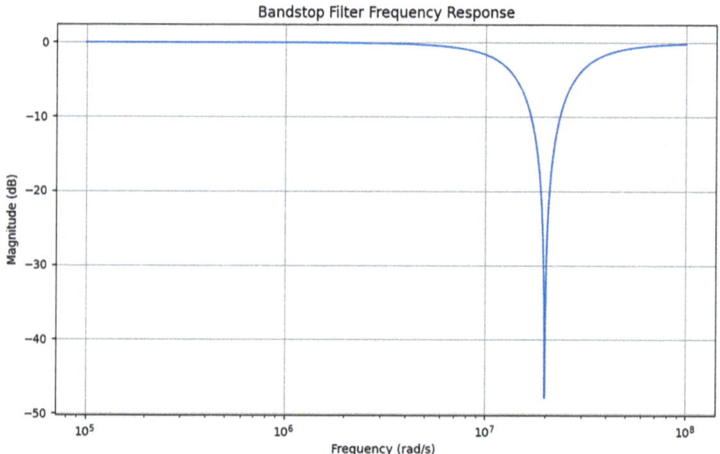

Fig. 16.6 The resulted bandstop filter frequency response

16.4 Designing an Array Antenna with Python (Colab)

Designing an array antenna in Python involves calculating the radiation pattern, beamwidth, and directivity of the array. Below is a step-by-step guide to designing a linear array antenna using Python in Google Colab.

Step 1: Import Required Libraries

```
import numpy as np
import matplotlib.pyplot as plt
```

Step 2: Define Array Parameters

Define the parameters of the array antenna, such as the number of elements, spacing, and wavelength.

```
# Array parameters
num_elements = 8 # Number of antenna elements
element_spacing = 0.5 # Spacing between elements (in wave-
lengths)
wavelength = 1 # Wavelength (normalized to 1 for simplici-
ty)
theta = np.linspace(-np.pi, np.pi, 360) # Angle range for
radiation pattern
```

Step 3: Calculate Array Factor

The array factor (AF) is a function of the angle and depends on the number of elements, spacing, and phase difference between elements.

```
def array_factor(num_elements, spacing, wavelength, theta,
phase_diff=0):
    k = 2 * np.pi / wavelength # Wave number
    d = spacing * wavelength # Distance between elements
    psi = k * d * np.cos(theta) + phase_diff # Phase dif-
ference
    af = np.sin(num_elements * psi / 2) / (num_elements *
np.sin(psi / 2))
    af = np.abs(af) # Magnitude of the array factor
    return af
# Calculate array factor
af = array_factor(num_elements, element_spacing, wave-
length, theta)
```

Step 4: Plot Radiation Pattern

Plot the radiation pattern of the array antenna.

```
# Plot radiation pattern
plt.figure(figsize=(10, 6))
plt.polar(theta, af, label='Array Factor')
plt.title('Radiation Pattern of Linear Array Antenna')
plt.legend(loc='upper right')
plt.grid(True)
plt.show()
```

Step 5: Calculate Beamwidth and Directivity

Calculate the half-power beamwidth (HPBW) and directivity of the array antenna.

```
# Find half-power beamwidth (HPBW)
half_power = np.max(af) / np.sqrt(2)
hp_indices = np.where(af >= half_power)[0]
hp_theta = theta[hp_indices]
hp_bw = np.max(hp_theta) - np.min(hp_theta)
print(f"Half-Power Beamwidth (HPBW):
{np.degrees(hp_bw):.2f} degrees")
# Calculate directivity
D = 2 * num_elements # Approximate directivity for a line-
ar array
print(f"Directivity: {D:.2f}")
```

Step 6: Add Steering Capability

To steer the beam in a specific direction, introduce a phase shift between elements.

```
# Beam steering
steering_angle = np.radians(30) # Desired steering angle
(30 degrees)
phase_diff = -2 * np.pi * element_spacing *
np.cos(steering_angle) # Phase difference
# Calculate steered array factor
af_steered = array_factor(num_elements, element_spacing,
wavelength, theta, phase_diff)
# Plot steered radiation pattern
plt.figure(figsize=(10, 6))
plt.polar(theta, af_steered, label='Steered Array Factor')
plt.title('Radiation Pattern with Beam Steering')
plt.legend(loc='upper right')
plt.grid(True)
plt.show()
```

Explanation of the code:

Array Factor: The array factor determines the radiation pattern of the array antenna.

Radiation Pattern: The plot shows how the antenna radiates energy in different directions.

Beamwidth: The angular width of the main lobe where the power is at least half of the maximum.

Directivity: A measure of how well the antenna focuses energy in a specific direction.

Beam Steering: By introducing a phase shift, the main beam can be steered in a desired direction.

This code provides a basic framework for designing and analyzing a linear array antenna. You can extend it to include more complex configurations, such as planar arrays or non-uniform spacing.

Full Code: Here is the full code for designing and analyzing a linear array antenna.

```
import numpy as np
import matplotlib.pyplot as plt

# Array parameters
num_elements = 8 # Number of antenna elements
element_spacing = 0.5 # Spacing between elements (in
wavelengths)
wavelength = 1 # Wavelength (normalized to 1 for
simplicity)
theta = np.linspace(-np.pi, np.pi, 360) # Angle range for
radiation pattern

# Array factor calculation
def array_factor(num_elements, spacing, wavelength,
theta, phase_diff=0):
    k = 2 * np.pi / wavelength # Wave number
    d = spacing * wavelength # Distance between elements
    psi = k * d * np.cos(theta) + phase_diff # Phase
difference
    af = np.sin(num_elements * psi / 2) / (num_elements *
np.sin(psi / 2))
    af = np.abs(af) # Magnitude of the array factor
    return af

# Calculate array factor
af = array_factor(num_elements, element_spacing,
wavelength, theta)

# Plot radiation pattern
plt.figure(figsize=(10, 6))
plt.polar(theta, af, label='Array Factor')
plt.title('Radiation Pattern of Linear Array Antenna')
plt.legend(loc='upper right')
plt.grid(True)
plt.show()
```

The radiation pattern of linear array antenna is illustrated in Fig. 16.7.

```
# Calculate half-power beamwidth (HPBW)
half_power = np.max(af) / np.sqrt(2)
hp_indices = np.where(af >= half_power)[0]
hp_theta = theta[hp_indices]
hp_bw = np.max(hp_theta) - np.min(hp_theta)
print(f"Half-Power Beamwidth (HPBW):
{np.degrees(hp_bw):.2f} degrees")
```

```
Half-Power Beamwidth (HPBW): 191.53 degrees
```

```
# Calculate directivity
D = 2 * num_elements # Approximate directivity for a
linear array
print(f"Directivity: {D:.2f}")
```

```
Directivity: 16.00
```

```
# Beam steering
steering_angle = np.radians(30) # Desired steering angle
(30 degrees)
phase_diff = -2 * np.pi * element_spacing *
np.cos(steering_angle) # Phase difference

# Calculate steered array factor
af_steered = array_factor(num_elements, element_spacing,
wavelength, theta, phase_diff)

# Plot steered radiation pattern
plt.figure(figsize=(10, 6))
plt.polar(theta, af_steered, label='Steered Array
Factor')
plt.title('Radiation Pattern with Beam Steering')
plt.legend(loc='upper right')
plt.grid(True)
plt.show()
```

The radiation pattern of linear array antenna with beam steering is shown in Fig. 16.8.

16.5 Calculating S-Parameters for an RF Network with Python (Colab)

Calculating scattering parameters (S-parameters) for an RF network involves solving the network's equations based on its topology and component values. Below is a Python code to calculate S-parameters for a simple 2-port RF network using the scikit-rf library, which is specifically designed for RF and microwave engineering.

Step 1: Install Required Libraries

First, install the scikit-rf library in Google Colab.

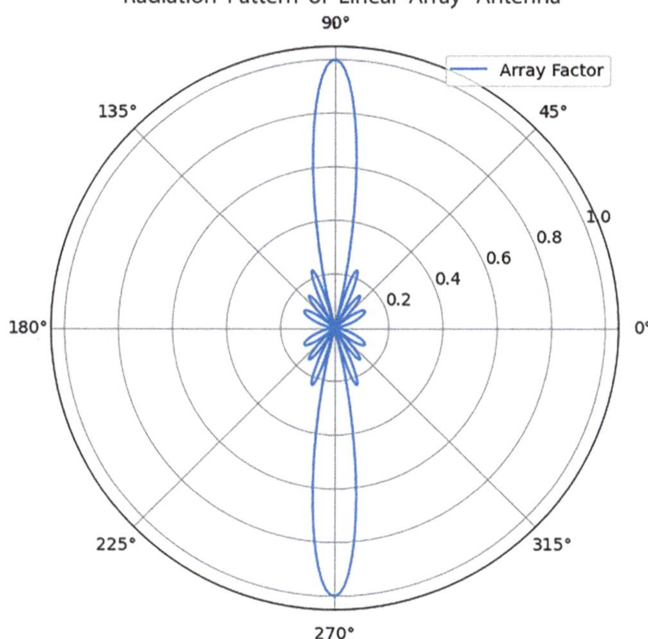

Fig. 16.7 The radiation pattern of linear array antenna

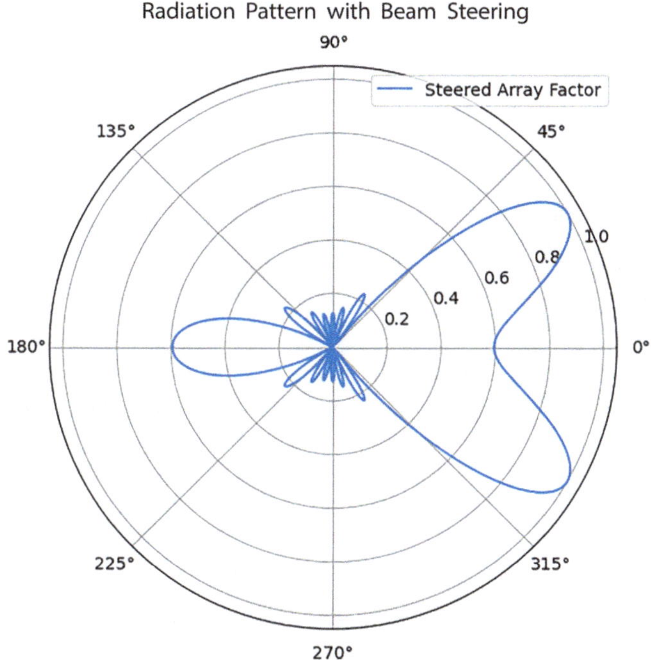

Fig. 16.8 The radiation pattern with beam steering

```
!pip install scikit-rf
```

Step 2: Import Required Libraries

```
import skrf as rf
import numpy as np
import matplotlib.pyplot as plt
```

Step 3: Define the RF Network

Define the RF network using its impedance matrix (Z-matrix) or admittance matrix (Y-matrix). For simplicity, let's consider a 2-port network with known impedance values.

```
# Define the frequency range
frequencies = np.linspace(1e9, 10e9, 100) # 1 GHz to 10
GHz
rf_freq = rf.Frequency.from_f(frequencies, unit='Hz')
# Define the impedance matrix (Z-matrix) for the 2-port
network
Z11 = 50 + 0j # Impedance at port 1
Z12 = 10 + 5j # Coupling impedance between port 1 and port
2
Z21 = 10 + 5j # Coupling impedance between port 2 and port
1
Z22 = 50 + 0j # Impedance at port 2
# Create the Z-matrix for all frequencies
# The shape of Z_matrix should be (n_frequencies, 2, 2)
Z_matrix = np.zeros((len(frequencies), 2, 2),
dtype=complex)
Z_matrix[:, 0, 0] = Z11  # Z11 is constant for all fre-
quencies
Z_matrix[:, 0, 1] = Z12  # Z12 is constant for all fre-
quencies
Z_matrix[:, 1, 0] = Z21  # Z21 is constant for all fre-
quencies
Z_matrix[:, 1, 1] = Z22  # Z22 is constant for all fre-
quencies
# Create a Network object from the Z-matrix
Z_network = rf.Network(z=Z_matrix, frequency=rf_freq)
```

Step 4: Convert Z-Matrix to S-Parameters

Convert the impedance matrix (Z-matrix) to scattering parameters (S-parameters).

```
# Convert Z-matrix to S-parameters
S_network = Z_network.s
```

Step 5: Plot S-Parameters

Plot the magnitude and phase of the S-parameters.

```
# Plot S11 (reflection coefficient at port 1)
plt.figure(figsize=(10, 6))
plt.subplot(2, 1, 1)
plt.plot(rf_freq.f_scaled,                    20              *
np.log10(np.abs(S_network.s[:, 0, 0])), label='S11 (dB)')
plt.title('S11 (Reflection Coefficient at Port 1)')
plt.xlabel('Frequency (GHz)')
plt.ylabel('Magnitude (dB)')
plt.grid(True)
plt.legend()
# Plot S21 (transmission coefficient from port 1 to port
2)
plt.subplot(2, 1, 2)
plt.plot(rf_freq.f_scaled,                    20              *
np.log10(np.abs(S_network.s[:, 1, 0])), label='S21 (dB)')
plt.title('S21 (Transmission Coefficient from Port 1 to
Port 2)')
plt.xlabel('Frequency (GHz)')
plt.ylabel('Magnitude (dB)')
plt.grid(True)
plt.legend()
# Print S-parameter matrix at a specific frequency
freq_index = 50  # Index of the frequency (e.g., 5 GHz)
print(f"S-parameter  matrix  at  {rf_freq.f[freq_index]  /
1e9:.2f} GHz:")
print(S_network[freq_index, :, :])  # Access S-parameters
directly
plt.tight_layout()
plt.show()
```

Step 6: Display S-Parameter Matrix

Display the S-parameter matrix at a specific frequency.

Explanation of the code:

Frequency Range: The frequency range is defined from 1 GHz to 10 GHz.

Impedance Matrix: The Z-matrix represents the impedance values of the 2-port network.

S-Parameters: The Z-matrix is converted to S-parameters using the scikit-rf library.

Plots: The magnitude of S11 (reflection coefficient) and S21 (transmission coefficient) are plotted in dB.

S-Parameter Matrix: The S-parameter matrix is displayed at a specific frequency.

This code provides a basic framework for calculating and analyzing S-parameters for an RF network. You can extend it to include more complex networks or additional parameters.

Full Code: Here is the full code for calculating and plotting S-parameters.

```
!pip install scikit-rf

Collecting scikit-rf
  Downloading scikit_rf-1.5.0-py3-none-any.whl.metadata
(6.8 kB)
Requirement already satisfied: numpy>=1.21 in
/usr/local/lib/python3.11/dist-packages (from scikit-rf)
(1.26.4)
Requirement already satisfied: scipy>=1.7 in
/usr/local/lib/python3.11/dist-packages (from scikit-rf)
(1.13.1)
Requirement already satisfied: pandas>=1.1 in
/usr/local/lib/python3.11/dist-packages (from scikit-rf)
(2.2.2)
Requirement already satisfied: typing-extensions in
/usr/local/lib/python3.11/dist-packages (from scikit-rf)
(4.12.2)
Requirement already satisfied: python-dateutil>=2.8.2 in
/usr/local/lib/python3.11/dist-packages (from
pandas>=1.1->scikit-rf) (2.8.2)
Requirement already satisfied: pytz>=2020.1 in
/usr/local/lib/python3.11/dist-packages (from
pandas>=1.1->scikit-rf) (2025.1)
Requirement already satisfied: tzdata>=2022.7 in
/usr/local/lib/python3.11/dist-packages (from
pandas>=1.1->scikit-rf) (2025.1)
Requirement already satisfied: six>=1.5 in
/usr/local/lib/python3.11/dist-packages (from python-
dateutil>=2.8.2->pandas>=1.1->scikit-rf) (1.17.0)
Downloading scikit_rf-1.5.0-py3-none-any.whl (3.5 MB)
────────────────────────────────────────────────────────

────────────────── 3.5/3.5 MB 37.7 MB/s eta 0:00:00

import skrf as rf
import numpy as np
import matplotlib.pyplot as plt

# Define the frequency range
frequencies = np.linspace(1e9, 10e9, 100)  # 1 GHz to 10
GHz
rf_freq = rf.Frequency.from_f(frequencies, unit='Hz')

# Define the impedance matrix (Z-matrix) for the 2-port
network
# For simplicity, we assume the Z-matrix is constant
across all frequencies
Z11 = 50 + 0j  # Impedance at port 1
Z12 = 10 + 5j  # Coupling impedance between port 1 and
port 2
Z21 = 10 + 5j  # Coupling impedance between port 2 and
port 1
```

```
Z22 = 50 + 0j  # Impedance at port 2

# Create the Z-matrix for all frequencies
# The shape of Z_matrix should be (n_frequencies, 2, 2)
Z_matrix = np.zeros((len(frequencies), 2, 2),
dtype=complex)
Z_matrix[:, 0, 0] = Z11  # Z11 is constant for all
frequencies
Z_matrix[:, 0, 1] = Z12  # Z12 is constant for all
frequencies
Z_matrix[:, 1, 0] = Z21  # Z21 is constant for all
frequencies
Z_matrix[:, 1, 1] = Z22  # Z22 is constant for all
frequencies

# Create a Network object from the Z-matrix
Z_network = rf.Network(z=Z_matrix, frequency=rf_freq)

# Convert Z-matrix to S-parameters
S_network = Z_network.s

# Plot S11 (reflection coefficient at port 1)
plt.figure(figsize=(10, 6))
plt.subplot(2, 1, 1)
plt.plot(rf_freq.f_scaled, 20 *
np.log10(np.abs(S_network[:, 0, 0])), label='S11 (dB)')
plt.title('S11 (Reflection Coefficient at Port 1)')
plt.xlabel('Frequency (GHz)')
plt.ylabel('Magnitude (dB)')
plt.grid(True)
plt.legend()

# Plot S21 (reflection coefficient at port 1)
plt.subplot(2, 1, 2)
plt.plot(rf_freq.f_scaled, 20 *
np.log10(np.abs(S_network[:, 1, 0])), label='S21 (dB)')
plt.title('S21 (Transmission Coefficient from Port 1 to
Port 2)')
plt.xlabel('Frequency (GHz)')
plt.ylabel('Magnitude (dB)')
plt.grid(True)
plt.legend()

# Print S-parameter matrix at a specific frequency
freq_index = 50  # Index of the frequency (e.g., 5 GHz)
print(f"S-parameter matrix at {rf_freq.f[freq_index] /
1e9:.2f} GHz:")
print(S_network[freq_index, :, :])  # Access S-parameters
directly

plt.tight_layout()
plt.show()
```

The S-parameter matrix at 5.55 GHz and frequency response of S11 as well as S21 are depicted in Fig. 16.9.

```
S-parameter matrix at 5.55 GHz:
[[-0.0074544 -0.01015067j  0.10023791+0.05138779j]
 [ 0.10023791+0.05138779j -0.0074544 -0.01015067j]]
```

16.6 Designing an RF L-Network for Impedance Matching with Python (Colab)

Below is a Python implementation for designing an RF L-Network for impedance matching of a complex load. This code can be run in Google Colab or any Python environment.

The Python code for RF L-Network design:

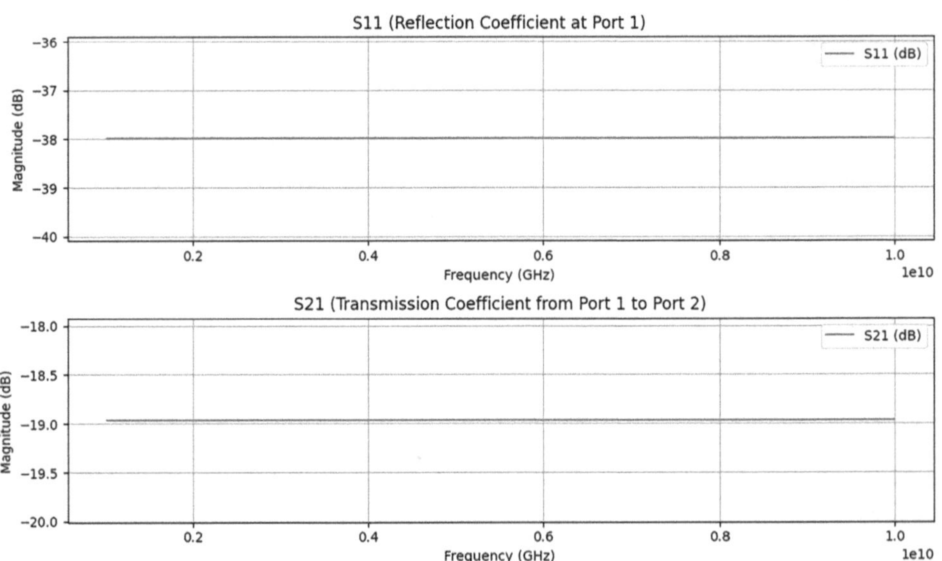

Fig. 16.9 The frequency response of S11 an S21

```
import numpy as np
# Constants
Z0 = 50  # Source impedance (typically 50 ohms)

# Input parameters
R_load = float(input("Enter load resistance (R_L in
ohms): "))  # Load resistance
X_load = float(input("Enter load reactance (X_L in ohms):
"))  # Load reactance
freq = float(input("Enter frequency (in Hz): "))  #
Operating frequency

# Complex load impedance
Z_load = R_load + 1j * X_load

# Normalized load impedance
z_load = Z_load / Z0
r_load = np.real(z_load)
x_load = np.imag(z_load)

# Check if the load is matchable with an L-Network
Q = np.sqrt(r_load * (1 + (x_load**2 / r_load)) - 1)
if Q <= 0:
    print("The load cannot be matched with an L-Network.
Choose a different topology.")
else:
    print(f"Q factor: {Q:.4f}")

# Calculate series inductor and shunt capacitor (High-
Pass configuration)
X_L = Z0 * Q  # Reactance of the series inductor
X_C = Z0 / Q  # Reactance of the shunt capacitor

# Calculate component values
L = X_L / (2 * np.pi * freq)  # Series inductor
C = 1 / (2 * np.pi * freq * X_C)  # Shunt capacitor


# Display results
print("\nL-Network Design (High-Pass Configuration):")
print(f"Series Inductor (L): {L * 1e9:.2f} nH")
print(f"Shunt Capacitor (C): {C * 1e12:.2f} pF")
# Calculate series capacitor and shunt inductor (Low-Pass
configuration)
X_C_series = Z0 / Q  # Reactance of the series capacitor
X_L_shunt = Z0 * Q  # Reactance of the shunt inductor

# Calculate component values
C_series = 1 / (2 * np.pi * freq * X_C_series)  # Series
capacitor
```

```
L_shunt = X_L_shunt / (2 * np.pi * freq)  # Shunt
inductor

# Display results
print("\nL-Network Design (Low-Pass Configuration):")
print(f"Series Capacitor (C): {C_series * 1e12:.2f} pF")
print(f"Shunt Inductor (L): {L_shunt * 1e9:.2f} nH")
```

How to Use the Code:

Open Google Colab or any Python environment.

Copy and paste the code into a new Python notebook or script.

Run the code.

Input the load resistance (RL), load reactance (XL), and operating frequency (f) when prompted.

The code will output the component values for both High-Pass and Low-Pass L-Network configurations.

Explanation of the code:

The code calculates the Q factor to ensure the load is matchable with an L-Network.

It computes the component values for both High-Pass (series inductor + shunt capacitor) and Low-Pass (series capacitor + shunt inductor) configurations.

The results are displayed in nanohenries (nH) for inductors and picofarads (pF) for capacitors.

This code is a practical tool for designing RF L-Networks for impedance matching. You can modify it to include additional features, such as plotting on a Smith Chart or simulating the network.

Here is the example for input and output of running the code.

```
Enter load resistance (R_L in ohms): 25
Enter load reactance (X_L in ohms): 50
Enter frequency (in Hz): 100e6
Q factor: 0.7071

L-Network Design (High-Pass Configuration):
Series Inductor (L): 56.27 nH
Shunt Capacitor (C): 22.51 pF

L-Network Design (Low-Pass Configuration):
Series Capacitor (C): 22.51 pF
Shunt Inductor (L): 56.27 nH
```

16.7 Conclusion

This chapter has introduced a transformative and practical methodology for modern radio frequency (RF) engineering by integrating artificial intelligence and cloud-based computation into the design workflow. We began by establishing the powerful synergy between DeepSeek, a state-of-the-art large language model (LLM) for code generation and problem-solving, and Google Colab, a cloud-based platform that eliminates the need for local software installation. This combination allows engineers to focus on high-level design concepts while automating the creation and execution of complex simulation scripts.

The core of the chapter demonstrated this new paradigm through a series of complete, executable Python implementations of fundamental RF components. We provided a step-by-step guide to building a Quadrature Amplitude Modulation (QAM) system, from binary data generation through modulation, demodulation, and verification. We systematically applied Python to the design of LC filters (low-pass, high-pass, bandpass, and band-stop), calculating component values and visualizing frequency responses. The chapter further extended this approach to linear array antenna design, enabling the calculation of radiation patterns, beamwidth, directivity, and electronic beam steering. We also showed how to calculate and visualize Scattering Parameters (S-Parameters) for a two-port network, and concluded with an interactive tool for designing L-Network impedance matching circuits.

In summary, this chapter has equipped you with a practical toolkit to harness computational power for RF design. Mastery of this workflow—leveraging AI for code generation and cloud platforms for execution—represents a significant evolution in the field. It empowers engineers to achieve rapid prototyping, perform rigorous analysis, and validate designs with unprecedented speed and flexibility, moving the discipline toward a more agile, code-centric, and automated future.

References

1. DeepSeek (2024) DeepSeek AI model: technical report and documentation. DeepSeek AI, Hangzhou
2. Google (2024) Google colaboratory: overview and documentation. Google LLC, California

Zeitfracht Medien GmbH
Ferdinand-Jühlke-Straße 7
99095 Erfurt, Deutschland
produktsicherheit@kolibri360.de